ENFORCING POLLUTION CONTROL REGULATION

Monitoring and enforcement issues must be analysed when determining the effectiveness of pollution control regulation, and clearly influence choices about how to regulate. This book demonstrates how an economic analysis of law enforcement can generate important insights into how best to enforce pollution control regulation. It seeks to provide a clear and accessible way into the law and economics literature on enforcement. More specifically, it uses Gary Becker's deterrence model which, by differentiating between two enforcement variables (namely the probability of apprehension and conviction and the severity of sanction), facilitates a comparison of the effectiveness of different enforcement tools in inducing desirable behaviour. As such, it provides a valuable analytical tool in considering how best to pursue cost-effective enforcement.

Major themes to be addressed include Becker's deterrence model and expansions thereof, reasons for compliance, environmental enforcement strategies and the importance of a deterrence threat and formal pollution control law enforcement mechanisms such as prosecution and criminal sanctions, administrative mechanisms and civil liability. The book argues that in pursuing cost-effective enforcement much can be learned from examining enforcement practices in different jurisdictions, and to this end the author examines pollution control laws, enforcement strategies and sanctions in Australia, Canada and England and Wales.

The book makes an important contribution to existing literature on environmental law enforcement, but its value extends beyond this. The theoretical framework adopted and the range of issues discussed make it of interest to regulatory and public law scholars more generally.

Enforcing Pollution Control Regulation

Strengthening Sanctions and Improving Deterrence

Carolyn Abbot

·HART·
PUBLISHING

OXFORD AND PORTLAND, OREGON
2009

Published in North America (US and Canada) by
Hart Publishing
c/o International Specialized Book Services
920 NE 58th Avenue, Suite 300
Portland, OR 97213-3786
USA
Tel: +1 503 287 3093 or toll-free: (1) 800 944 6190
Fax: +1 503 280 8832
E-mail: orders@isbs.com
Website: http://www.isbs.com

Hart Publishing Ltd, 16C Worcester Place, Oxford, OX1 2JW
Telephone: +44 (0)1865 517530 Fax: +44 (0)1865 510710
E-mail: mail@hartpub.co.uk
Website: http://www.hartpub.co.uk

British Library Cataloguing in Publication Data
Data Available

ISBN: 978-1-84113-925-8

Typeset by Hope Services, Abingdon
Printed and bound in Great Britain by
Lightning Source UK Ltd

CONTENTS

TABLE OF CASES

AUSTRALIA

CANADA

NEW ZEALAND

UNITED KINGDOM

UNITED STATES

TABLE OF LEGISLATION

AUSTRALIA

Commonwealth

New South Wales

Alberta

British Columbia

Ontario

Quebec

UNITED KINGDOM

EUROPEAN COMMUNITY LEGISLATION

1

Environment, Enforcement and Deterrence

I Introduction

I CANNOT RECALL the last time I read a newspaper or even listened to the news on the radio without coming across an 'environment-related' issue. Whether it be an announcement by my local authority that a new waste recycling collection service is to be introduced in Stockport, the unveiling of plans to expand Heathrow airport or an inside story on the unlawful destruction of the African rainforests, we are constantly being reminded of the impact that human activities are having on the natural environment. Public interest groups and the media have gone far in bringing environmental issues to the forefront of our minds and will no doubt have prompted some of us to be more environmentally conscious. But this is not enough. Governmental action at international, national and local level plays a key role in influencing behaviour and, since the 1960s, industrialised nations have witnessed an unprecedented growth in the use of legal and economic instruments designed both to prevent and minimise the risk of environmental harm occurring and to mitigate existing environmental damage. In fact, many countries worldwide now have an impressive array of laws aimed at controlling potentially polluting activities. This monograph is not, however, concerned with the necessity and desirability of environmental regulation, or with its substantive content. These are issues dealt with comprehensively elsewhere.[1] Rather, this book focuses on regulatory enforcement, a complex and at times controversial topic. On a day to day basis, enforcement officials face a number of challenges, all of which hamper their quest for effective and efficient enforcement:

> Resources are often thinly spread and errant behaviour is difficult to detect. Regulatory objectives are not always clear and legal powers may be limited. Enforcement functions are often distributed across numbers of regulators who struggle to co-ordinate their activities. Further, it is often extremely hard to measure the success or failure of

[1] For an excellent introduction to questions such as why and how to regulate see C Hilson, *Regulating Pollution: A UK and EC Perspective* (Oxford, Hart Publishing, 2000).

regulation. Even if such measurement is possible, it may be very difficult to improve the regulatory system by adjusting enforcement strategies and legal powers.[2]

It is unsurprising, then, to discover that socio-legal scholars, economists, criminologists, sociologists and political scientists (to name but a few) have concerned themselves for some time with how best to enforce the law. This rich diversity of literature provides a stimulating backdrop against which regulatory enforcement can be analysed.

This monograph demonstrates how an economic analysis of law enforcement can generate important insights into how best to enforce pollution control regulation, namely regulation that limits the discharge of environmental contaminants to air, water and land. Although I make no claim to original economic analysis, I do seek to provide a clear and accessible way into the law and economics literature on enforcement, more specifically, Gary Becker's deterrence hypothesis.[3] An expanded version of the Becker theory is used to analyse the cost-effectiveness of different sanctions and penalties in securing a high level of compliance, with a view to determining whether the deterrent effect of enforcement action can be improved by strengthening the sanctioning toolkit.[4] It is not necessary to utilise and understand complex mathematical models, equations or graphs in order to benefit from taking a law and economics approach.

Granted, the economic approach is by no means the only basis on which to consider enforcement issues and there are, as I shall discuss in Chapter 2, a number of difficulties inherent in using this approach.[5] But by differentiating between two enforcement variables (namely the probability of apprehension and conviction and the severity of sanction), it allows us to compare the effectiveness of different enforcement tools in inducing desirable behaviour and as such, provides a valuable analytical tool in the study of law and enforcement. By focusing on sanctions and penalties, I emphasise the role of formal, legal enforcement mechanisms such as criminal prosecution, administrative notices and monetary administrative penalties.[6] This is not to detract from the importance of informal enforcement tools such as education and persuasion, which are indeed a vital component of the enforcement toolkit. Rather, the presence of a strong and credible deterrence threat actually facilitates and enables a less sanction-based approach to enforcement.

My analysis is discussed taking examples of pollution control enforcement strategies and sanctions in Australia, Canada and England and Wales. As Australia and Canada are federal systems, the power to make law is divided between the federal government and the government of states or provinces. In addition to federal controls, there are therefore separate, state-specific regulations and independent institutional arrangements aimed at securing compliance. With this in mind, and in

[2] R Baldwin and J Black, 'Really Responsive Regulation' (2008) 71 *MLR* 59, 59.

[3] GS Becker, 'Crime and Punishment: An Economic Approach' (1968) 76 *Journal of Political Economy* 169.

[4] For an introduction to cost-effectiveness analysis, see below s II C ii.

[5] For detail, see ch 2 s III.

[6] For detail, see chs 7 and 8.

order to make my task more manageable, I focus on pollution control regulation and its enforcement at federal level, and within one or two states or provinces. More specifically, in Australia I focus on enforcement at the Commonwealth level and in the states of New South Wales and Victoria. In Canada, I consider federal enforcement and enforcement in the province of Ontario. These countries and states were chosen for a number of reasons. First, in all three countries, the command and control, licensing model forms the backbone of pollution control regulation.[7] Criminal liability and sanctions underpin the regulatory framework, and a variety of other formal sanctions, both administrative and civil, are available to enforcement agencies. Second, they have all introduced (or are in the process of introducing) interesting and innovative sanctions that can respond to instances of non-compliance and potentially strengthen the deterrence impact of enforcement. Third (and broadly speaking), they adopt a similar approach to enforcement. Unlike the US Environmental Protection Authority which is relatively confrontational and coercive, environmental regulators in Australia, Canada, and England and Wales take a far more persuasive and educative approach to enforcement although a compliance-based strategy, which sees prosecution as a 'last resort', still relies heavily on the presence of a real and effective deterrent threat. Finally, the states and provinces of New South Wales, Victoria and Ontario have been singled out as highly populated and industrialised areas with a history of industrial pollution.

II Regulation: Compliance, Enforcement and Penalties

Before introducing the deterrence hypothesis, it is important to define the scope of this study and to clarify key concepts. This section duly begins with an overview of regulatory techniques, focusing on command and control, an instrument which has traditionally been used to regulate emissions from industrial processes. It then considers, in Sections B, C and D, the meaning of three important concepts, namely compliance, enforcement, and penalties and sanctions. It examines, from a theoretical perspective, how much compliance and enforcement is desirable, and introduces cost-effectiveness analysis as a method of assessing different enforcement strategies and penalties. As is implicit in the title, this monograph considers enforcement from a public or regulatory perspective. This section concludes by introducing the public versus private enforcement debate.

A Regulatory Techniques

The environmental law toolbox contains a range of regulatory instruments or techniques.[8] Most jurisdictions have traditionally relied on the use of direct

[7] For a brief discussion of 'command and control' regulation see s II A below.
[8] For an overview of regulatory techniques see Hilson above n 1, ch 6 .

regulation, commonly referred to as 'command and control'. Although the term 'command and control' embraces a range of regulatory techniques, there are two basic characteristics to any such system. First, a government authority will prescribe environmental quality standards or targets which must be complied with (the 'command'). These standards or targets may, for example, require the use of specific technology in an industrial process or set maximum amounts of a pollutant that can be given off by a plant or other source. Second, non-compliance with these standards or targets will attract a negative sanction (the 'control') such as prosecution in the criminal courts. In many instances, 'command and control' regulation is found in conjunction with the use of prior approval mechanisms such as licences, authorisations or permits. Under a system of prior approval (a form of *ex ante* as opposed to *ex post* control), it is unlawful to carry out certain activities without a licence. When setting licence conditions, environmental quality standards can be drafted so as to reflect the local environment and the heterogeneity of regulated firms. This 'command and control' technique is of course by no means the only environmental policy instrument that is appropriate for regulating the environment. Market-based or economic instruments such as taxes, subsidies and trading schemes are being used to tackle issues such as greenhouse gas emissions, waste production and disposal and habitat conservation. Eco-labelling and impact assessment are important in ensuring the disclosure of environmental information. And voluntary instruments including environmental management systems, environmental auditing and environmental reporting can, amongst other things, provide a flexible approach to meeting environmental targets. 'Command and control' has, however, been used for decades to regulate more traditional forms of industrial pollution.

That command and control regulation has had some success in tackling polluting behaviour is undisputed.[9] But it is not without its critics. One perennial challenge relates to the problem of enforcement. In the absence of voluntary compliance, effective enforcement is absolutely essential if the intended purpose of the regulatory regime is to be achieved. In other words, it is the enforcement process which puts the flesh on the bones of any regulatory system. Regulatory authorities are, in the main, responsible for ensuring compliance with environmental regulation, and officials can utilise a range of enforcement styles and tools in promoting compliance with the law.[10] In some cases, regulators may adopt a legalistic, deterrence-based approach to enforcement, which relies on the use of formal legal sanctions such as criminal prosecution in cases of non-compliance. In other instances, a persuasive, educative style may be preferred to coax individuals and companies into compliance. Alternatively, a mix of approaches may better respond to the drivers and barriers which explain non-compliance. Whatever the enforcement strategy, it is clear that monitoring and enforcement issues must be analysed when determining the

[9] For a summary see C Abbot, 'Environmental Command Regulation' in B Richardson and S Wood (eds), *Environmental Law for Sustainability* (Oxford, Hart Publishing, 2006).

[10] For theories of enforcement, see ch 3. See chs 6, 7 and 8 for an analysis of enforcement tools available to pollution control regulators in parts of Australia and Canada, and England and Wales.

effectiveness of environmental regulation and clearly influence choices about how to regulate. My discussion of enforcement occurs in the context of command and control regulation, specifically as applied to industrial pollution. This should not, however, detract from the fact that adopting a law and economics approach enables us to develop our understanding of the enforcement process more generally for other regulatory techniques such as economic instruments and informational measures all need enforcing if they are to achieve their goals.

B Compliance

i Compliance with What?

In the regulation literature, a distinction is sometimes made between the need to ensure compliance with the collective goals of a regulatory scheme (such as reducing pollution generally or minimising the emission of a particular substance into the air), and compliance with specific regulatory standards. However, as noted by Yeung, 'the tendency of some theorists to use the term interchangeably and inconsistently' can lead to confusion.[11] Compliance with, on the one hand, collective goals and, on the other, regulatory standards, may also not necessarily be consistent. It may be the case that, for example, regulatory standards are so poorly designed that compliance with these standards does not deliver on the collective goals of the regulatory system.[12] For the sake of clarity, I refer to 'rule compliance', that is compliance with regulatory standards, as this is most commonly associated with the regulatory enforcement process.

ii Desirable Levels of Compliance

How much compliance are regulators looking to achieve? Is absolute or perfect compliance with the law possible or indeed desirable? Much of the law and economics literature on enforcement promotes optimal compliance, where the social benefits of compliance are equal to the social costs of achieving that level of compliance. There are, however, difficulties in using this approach and so I utilise a different normative goal for enforcement agencies, namely the achievement of a high level of compliance. This section considers the merits of perfect, optimal and high levels of compliance.

There could be a strong public interest case in favour of perfect compliance with pollution control regulation. The increasing availability of scientific and technical evidence on the negative impact of potentially polluting substances on the environment has led to considerable support for some form of government intervention. Perfect compliance would surely then be a desirable policy goal for if the regulated

[11] K Yeung, *Securing Compliance : A Principled Approach* (Oxford, Hart Publishing, 2004) 11.
[12] Additional examples are provided by Yeung, *Ibid.*

community were to comply fully with its environmental obligations, then environmental degradation would be minimised, at least to the extent envisaged by the regulations. Where non-compliance with regulatory standards leads directly to significant environmental, health or property damage, few would disagree. But from an economic perspective, perfect compliance is neither practicable nor desirable.[13]

Why? In answering this question, it is important to acknowledge that when deciding how much compliance is economically desirable, three key factors must be taken into account. First, regulatory offences give rise to losses. The commission of a substantive 'pollution' offence (as opposed to an administrative offence, such as failure to supply a regulatory authority with prescribed information) will either cause harm or at least increase the possibility of harm. Pollution damage can be catastrophic, may extend across geographical boundaries and can potentially affect future generations. In many instances, remediation of the damage caused by the pollution is difficult, costly and maybe even impossible. The social benefit of compliance is therefore the avoidance of such harm. Second, achieving compliance imposes costs on the regulated community.[14] The 'direct' costs of pursuing compliance (and thus, the costs of avoiding any harm that the law was designed to prevent) can be high.[15] Such costs include the capital expenditure on new equipment, the adaptation of old plant, additional annual operating costs and productivity losses.[16] Administrative costs associated with the permit application procedure, record keeping, acquiring relevant information and perhaps negotiating with the enforcement agency must also be factored into the calculation.[17] And third, regulations have to be enforced if they are going to work.[18] The enforcing

[13] This argument is made throughout the economic literature on enforcement. See, eg, G Stigler, 'The Optimum Enforcement of Law' (1970) 78 *Journal of Political Economy* 526.

[14] Industry has, for countless decades, bemoaned the unreasonable costs imposed on them by irrationally designed and unfairly enforced environmental regulations, although 'there seems considerable evidence that industry and trade associations regularly exaggerate the cost of implementing environmental regulations.' House of Commons Environmental Audit Committee 7th Report, *Pre-Budget 2004 and Budget 2005: Tax, Appraisal, and the Environment* HC (13 April 2005) 261 [67]. Also see International Chemical Secretariat Report 6:04, *Cry Wolf—Predicted Costs by Industry in the Face of New Regulations* (April 2004).

[15] There are also often indirect or unintended consequences of regulation, which may have a huge impact over and above any examination of compliance costs. Broadly speaking, the indirect costs of regulation fall into three categories: allocative inefficiency, productive inefficiency, and the inhibition of technical change.

[16] Although environmental standards often leave a company free to select methods of pollution reduction that are most cost-effective and appropriate in their particular circumstances, regulated firms nevertheless have strong incentives to adopt the particular technology underlying the standard because its use will readily persuade regulators of compliance. The adoption of such technology may subsequently drive up compliance expenditure. Moreover, compliance costs are generally higher than expected because regulated entities will tend to over-comply in order to avoid any risk of enforcement action. From a societal perspective therefore, marginal compliance, whereby the regulated community uses just the amount of resources necessary to remain within the legal limits, may be most efficient.

[17] And with the advent of full cost recovery charging, the majority of administrative costs incurred by the regulatory authority are subsumed by the regulated companies.

[18] Note that 'rule-making' costs, that is the costs associated with designing and implementing legal standards could also be factored into this analysis. But since this chapter is concerned with day-to-day enforcement and rule-making costs have already been incurred, they will not affect an enforcement

authority will incur costs in identifying breaches and taking appropriate enforcement action, whether informal or formal, administrative, civil or criminal. In any given situation, the amount of enforcement therefore depends, to a great extent, on the amount of resources devoted to the task. It is conceivable that an environmental regulator, with a large and well-financed enforcement branch, could identify and apprehend all those who, for example, dispose of waste unlawfully. But the costs inherent in pursuing such an aggressive enforcement strategy make this approach unworkable and so society normally gives enforcement agencies a budget, thereby dictating a much lower level of enforcement.

In deciding how much compliance (and therefore enforcement) is economically desirable, it is necessary to compare these three factors, that is the avoided social losses (ie harm), with compliance and enforcement costs.[19] In some cases, an offence may cost more to avoid, in terms of compliance and enforcement costs, than the harm that would occur in the event of non-compliance. Compliance in cases such as these is economically undesirable, as ultimately it would impose greater costs on society and would not advance society's well-being. However, where the harm that would be avoided as a result of compliance is more than compliance and enforcement costs, then compliance would be economically desirable. It is therefore necessary to balance compliance and enforcement costs against the harm that would be avoided. On this basis, much of the economic literature on regulatory enforcement concludes that the key to an efficient enforcement strategy is to ensure cost-justified or optimal compliance, achieved when the marginal social benefits accruing from compliance (avoided harm) are equal to the marginal costs incurred in achieving that level of compliance (compliance and enforcement costs).[20]

So, on the basis of our discussion so far, from an enforcement perspective the goal of a regulatory authority would be optimal enforcement, achieved where the costs of enforcement are outweighed by the benefits (in terms of reduced violations and avoidance of harm). However, optimal compliance and enforcement are difficult to put into practice. One problem relates to the quantification of costs and benefits which is made more complex when, as in our case, some of the resources which must be valued (such as the environment) are not traded. So, rather than determine what is optimal in terms of compliance and enforcement, one can adopt a less ambitious mode of economic appraisal as a means of evaluating enforcement polices and practices. I adopt an alternative normative goal of enforcement, namely enforcement which achieves a 'high level of compliance'. In other words, I examine the extent to which enforcement can achieve a specified benefit; that is a high level of compliance.

officer's decision and will be ignored: CG Veljanovski, 'The Economics of Regulatory Enforcement' in K Hawkins and J Thomas (eds), *Enforcing Regulation* (Boston, Kluwer-Nijhoff Publishing, 1984).

[19] These factors are not dissimilar to Calabresi's characterisation of accident costs as primary, secondary and tertiary. According to Calabresi, the goal of accident law and policy should be to bring about the socially preferred accident-cost level ie to optimise accident costs. See generally G Calabresi, *The Costs of Accidents: A Legal and Economic Analysis* (New Haven, Yale University Press, 1970).

[20] Above n 13.

C Enforcement

i Defining Enforcement

A useful starting point in defining enforcement is to consider it as one stage in the process of regulatory implementation. Yeung argues that once the collective goals underpinning a regulatory programme have been identified, regulatory implementation is a three-staged process involving 'institutional design, standard setting and the monitoring and enforcement of regulatory standards.'[21] The first institutional design stage includes designing both the institutions and mechanisms to be used as a means of regulatory implementation. It may therefore require a consideration of whether or not to utilise, for example, economic instruments, voluntary mechanisms or licensing schemes as a means of furthering regulatory objectives. Standard setting, the second stage, is most obviously relevant in a command and control system whereby rules, offences and sanctions must be determined although in the case of other regulatory instruments such as economic incentives, the rules of engagement also have to be established before a system can become operable. Regulatory enforcement is not, however, concerned with either of these two stages. In other words, the enforcement process begins when the parameters of legal control, both general and specific, have been established.

The final stage in Yeung's taxonomy of regulatory implementation is monitoring and enforcement. Monitoring requires regulators to oversee the extent to which the regulated community is in compliance and will usually involve some level of inspection. Regulators will have their own inspection regimes which may inter alia focus on responding to accidents or threats, identifying targets according to risk or indeed inspecting all or a set proportion of businesses. Enforcement relates to the investigation of an incident in order to establish the nature and extent of non-compliance, and the taking of action to correct non-compliance. The Department of Environment, Food and Rural Affairs (DEFRA) in England and Wales has identified the four key components of enforcement as:

—direct dealings between the regulator and those with statutory responsibilities ('the operator');
—an investigation sufficient to establish the nature and extent of non-compliance;
—a range of measures intended to prevent and correct non-compliance with environmental requirements;
—ensuring that operators take short-term or long-term corrective action.[22]

The process of investigation will be constrained by policies which guide regulators in deciding how and when an investigation should take place, and rules which determine the legal boundaries of regulatory powers to, for example, enter

[21] Above n 11, 11–12.
[22] DEFRA, *Review of Enforcement in Environmental Regulation, Report of Conclusions* (October 2006) 7–8.

premises and seize property. Enforcement is often equated with prosecution, a formal legal process in which the courts impose a sanction for violating the law. But it is about more than that. It includes the use of formal, administrative mechanisms such as the serving of administrative notices, the revoking of licences and the imposition of monetary administrative penalties, and possibly the launch of formal civil proceedings with a view to inter alia, ceasing an unlawful activity or claiming compensation for harm done. Informal techniques such as advice, negotiation, education and persuasion are also of central importance in the enforcement process. As noted by Hutter, 'these [informal enforcement techniques] were used by all law enforcement officials, but came into particular prominence in the regulatory arena.'[23] All these mechanisms, from persuasion and education, to prosecution and licence revocation are important in enforcing regulation.

ii Enforcement, Compliance and Cost-effectiveness Analysis

As discussed previously, this monograph proceeds on the basis that the goal of a regulatory agency is to promote a high level of compliance with regulatory standards. It then applies what is known as cost-effectiveness analysis to that given benefit. Cost-effectiveness analysis 'refers to the evaluation of alternatives according to both their costs and their effects with regard to producing some outcome.'[24] According to Ogus, it has two principal functions. First, it can be used 'to determine the type of intervention that will maximise benefits for a given level of costs, as specified by policy makers.'[25] Second, it may ascertain the type of intervention which will produce specified benefits at the lowest cost. As benefits are usually more difficult to quantify than costs, the second of these two functions is utilised more frequently and therefore forms the basis of my analysis.

As a tool of economic analysis, cost-effectiveness is often encountered when evaluating the most appropriate type of regulatory intervention. So, for example, it has been used to compare licensing techniques and economic instruments as modes of regulation. The most cost-effective regulatory option will be that which achieves a given regulatory objective at the lowest cost. I, however, am using cost-effectiveness analysis as the basis of an investigation into the public enforcement of pollution control regulation. More specifically, I will be exploring how different enforcement policies and tools can be used to achieve a desired behavioural outcome, namely compliance with the law. In short, the objective of my study is to investigate which enforcement policies and tools are the most cost-effective in achieving a high level of compliance with pollution control regulation. So, having established the objective of the analysis, I need to develop a theoretical framework

[23] B Hutter, *Compliance: Regulation and Environment* (Oxford, Clarendon Press, 1997) 14.

[24] HM Levin and PJ McEwan, *Cost-Effectiveness Analysis*, 2nd edn (California, Sage Publications, 2001) 10.

[25] AI Ogus, *Costs and Cautionary Tales—Economic Insights for the Law* (Oxford, Hart Publishing, 2006) 290.

against which we can analyse the cost-effectiveness of different enforcement options. In section III, I present this framework, which is based on a simple model of deterrence.

D Penalties and Sanctions

In adopting an analytical framework which utilises a deterrence hypothesis, I focus on the use of formal 'penalties' or 'sanctions', terms which may be and are used interchangeably to denote responses to regulatory infractions.[26] An aim of all penalties is the deterrence of wrongdoing, both specific (deterring the person sanctioned from repeating the offence) and general (deterring others from engaging in the unlawful behaviour). Although penalties also have a set of more narrowly defined aims, categorised by the Australian Law Reform Commission as retribution (or 'just deserts'), social condemnation (expressed through the stigma association with criminal conviction or a severe penalty), protection of third parties or the public generally (applied particularly through the use of administrative penalties such as licence suspension or revocation) and payment of reparation or compensation, deterrence is ultimately an aim of all penalty regimes.[27] I limit my analysis to criminal prosecution and civil proceedings[28] (and the penalties consequent to a successful action) and formal administrative sanctions (that is, penalties that can be imposed by a regulator without recourse to a court or tribunal) such as administrative notices, administrative monetary penalties and licence suspension/revocation. This is not to detract from the fundamental importance of informal, persuasive techniques, which are undoubtedly pervasive throughout the enforcement process. However, whilst they arguably impose costs on regulated entities and possibly raise the likelihood of formal action in the event of future non-compliance, they less obviously display the deterrence-based characteristics of penalties.

[26] The Australian Law Reform Commission (ALRC), in its report on *Principled Regulation*, examined the distinction between penalties and sanctions more closely, and concluded that the two terms have slightly different meanings. The term 'penalty' tends to carry with it negative connotations of punishment, which although traditionally meted out by the criminal law, is now more widely accepted to include a penalty imposed by a civil or administrative process. The term 'sanction' can include both positive and negative persuasion, and as such is broader in remit than a penalty. The ALRC concludes by defining 'penalty' broadly to include the withholding of benefits of entitlements. Australian Law Reform Commission 95, *Principled Regulation: Federal Civil and Administrative Penalties in Australia* (2002) ch 2, paras 2.25–2.39.

[27] Australian Law Reform Commission Discussion Paper 65, *Securing Compliance: Civil and Administrative Penalties in Australian Federal Regulation* (April 2002) ch 3, para 3.5. Penological approaches to penalties are discussed in more detail in ch 7 s II B.

[28] In some jurisdictions, regulators can bring a civil action for an injunction and/or damages for breach of regulatory obligations. Civil proceedings are discussed only briefly in ch 7 s VII.

E Public v Private Enforcement

By referring to 'regulatory enforcement' in the title of this monograph, it is clear that my investigation focuses on the public enforcement of pollution control regulation, that is, enforcement by a public agency, as opposed to the enforcement of government regulation by private parties. This is not to deny the role that private enforcement can play in securing compliance with both collective goals and regulatory standards. In fact, private parties make an extremely important contribution to the enforcement effort and it is worth pausing at this juncture to consider briefly why this is the case. Of course, private parties can enforce private, individual rights through common law civil actions such as negligence and nuisance. But this is not my concern. Nor am I looking at the policy of many environmental regulators to co-opt private individuals into the enforcement process by encouraging them, for example, to report suspected violations. Although the information they supply may result in legal intervention, the decision to enforce rests with the public enforcer, as does the cost and resource implications associated with action. In this vein, private parties have a clear, indirect role, in public enforcement. Rather, I wish to consider briefly here the role that private parties can and should play in directly enforcing public regulation by for example, initiating a private prosecution or seeking a court-imposed remedy for injury caused by those who breach regulatory standards.[29]

Private law enforcement is, in fact, a pervasive feature of existing systems of environmental regulation. In the UK, for example, third parties such as public interest groups can bring criminal proceedings against an individual or company that breaches regulatory controls.[30] The Royal Society for the Prevention of Cruelty to Animals (the RSPCA) has been particularly prominent in prosecuting individuals for animal welfare offences, and, in 2007, 1,860 cases were reported to its prosecutions department, resulting in the conviction of 1,104 defendants.[31] Private enforcement is supported by economists on the basis that private individuals have a greater incentive to arrive at an efficient (optimal) level of enforcement. Unlike public enforcers, private individuals internalise both the costs and benefits of enforcement.[32] This has led economists such as Becker and Stigler to advocate a system in which all enforcement, including the criminal law, is privatised.[33]

[29] See GS Becker and GJ Stigler, 'Law Enforcement, Malfeasance, and the Compensation of Enforcers' (1974) 3 *Journal of Legal Studies* 1, and the response by W Landes and RA Posner, 'The Private Enforcement of Law' (1975) 4 *Journal of Legal Studies* 1.

[30] This right exists at common law and is preserved by Prosecution of Offences Act 1985 s 6(1).

[31] See <http://www.rspca.org.uk/servlet/Satellite?pagename=RSPCA/RSPCARedirect&pg=prosecutions&marker=1&articleId=1172248241558> accessed 5 January 2009.

[32] Yeager, eg, notes that public enforcement agents in the United States have been able to bring a larger number of enforcement actions then regulatory agencies, at a fraction of the cost. See PC Yeager, *The Limits of Law: The Public Regulation of Private Pollution* (Cambridge, CUP, 1991) 321. See also W Landes and RA Posner (above n 29) and AM Polinsky, 'Private versus Public Enforcement of Fines' (1980) 9 *Journal of Legal Studies* 105.

[33] Above n 29.

Although Becker and Stigler's model has been refined to reflect the fact that private enforcement will not always be desirable and more efficient[34] there is no doubt that, regardless of its efficiency claims, it acts as an important complement to public enforcement efforts.[35] For example, it can fill the unavoidable 'enforcement gap' created by under-resourced public authorities, it can reinforce social values and send a signal to the regulated community that certain behaviour is not tolerated, and it can enhance accountability by providing a check and balance on regulatory action (or inaction).

However, to suggest that private enforcement is unproblematic would be naïve.[36] It should not and cannot replace regulatory enforcement efforts. There is always a risk, for example, that private litigation will be directed not at the most serious offenders but at those most likely to settle out of court. Furthermore, effective enforcement takes place, as we shall see, in the context of an ongoing relationship between regulator and regulated in which a mix of co-operative and deterrence-based strategies are used to determine a suitable outcome. Private enforcement may therefore 'disrupt what might otherwise be a workable regulatory arrangement.'[37] Moreover, private enforcement has the potential to lead to inconsistency and unpredictability, neither of which is desirable. These general concerns do not and should not detract from the role that it can play in promoting compliance. Rather, they emphasise the need for private enforcement to complement but not replace public enforcement, the challenge being:

> [T]o harness the constructive elements of private enforcement, neutralize those elements which might be counterproductive, and combine them systematically with public enforcement.[38]

Recent work in this area has in fact recognised that regulatory enforcement does not take place in a vacuum, but operates within a complex regulatory environment. This 'regulatory space' is occupied not just by public regulators or indeed public interest groups and civil society, but by other third parties including trade associations, professionals, and insurance companies, all of which can support the public enforcement of regulatory schemes. Grabosky, for example, promotes the use of regulatory surrogates to fill the gap created by the under-resourcing of public enforcement efforts.[39] Instruments, such as record-keeping and disclosure, co-option of private interests, incentives for third parties for the co-production of law enforcement services, contracting out and abdication, will, he argues, assist regulatory and criminal justice authorities with their law enforcement activity, either through the provision of information or services (which will not only

[34] Landes and Posner above n 29.

[35] See further K Yeung, 'Privatising Competition Regulation' (1998) 18 *OJLS* 581, 586–91.

[36] For a summary, see, eg, N Gunningham and P Grabosky, *Smart Regulation: Designing Environmental Policy* (Oxford, OUP, 1998) ch 3, 105–06. Also see Yeung, *Ibid* 591–2.

[37] *Ibid* (Gunningham and Grabosky) 105.

[38] *Ibid* 106.

[39] P Grabosky, 'Using Non-Governmental Resources to Foster Regulatory Compliance' (1995) 8 *Governance* 527.

inform regulatory enforcement strategy and action but will also increase the regulated entity's perception of the risk of being caught and convicted). Although the co-option of third party interests does not come without its risks, many of these instruments have the potential to make state regulatory enforcement activity more effective and efficient.

III Introducing the Deterrence Hypothesis[40]

In 1968, Gary Becker, an eminent Chicagoan lawyer-economist and Nobel prize winner[41] published a controversial and yet hugely influential paper on the economic analysis of the criminal law. Stated simply, Becker's deterrence hypothesis examines how the law, as an external inducement, can influence behaviour. It is based on the premise that firms and individuals are rational economic beings who are motivated by wealth or utility-maximising intentions, an economic postulate which is most obviously relevant to those engaging in market-based transactions. Becker, however, believed that this reasoning was also applicable in areas such as criminal and family law. Using his analytical framework, individuals and firms are discouraged from engaging in non-compliant behaviour by virtue of the fact that it may be detected and punished.

Let's consider an example. Recently, I had to call in at my local Post Office Delivery Depot to collect a parcel. I was on my way to work and had a nine o'clock contract lecture to prepare for. Parking at the depot is restricted to three spaces, all of which were full when I arrived. Bearing in mind that the nearest car park is a couple of minutes away, I decided to park (illegally) on a double yellow line on the road next to the depot. My decision may have resulted from my intuitive calculation that the costs and consequences of being caught did not exceed the benefit or utility I gained in saving time by not driving to the car park. Now of course, social scientists such as social psychologists would focus more on the internal reasons for non-compliance—perhaps that I did not feel any moral compulsion to obey parking laws.[42] This may have been the case. But as acknowledged by Ogus:

> [I]f it be accepted—which typically is the case—that external inducements do play a significant part in motivating behaviour, then, for legal policy purposes, it makes sense to analyse such phenomena in depth because influencing the mind-set, which makes up the 'internal' response, is invariably more costly and more difficult to procure than altering the operation of the external inducements.[43]

[40] All the issues raised in this section are discussed in detail in ch 2.
[41] RA Posner, 'Gary Becker's Contribution to Law and Economics' (1993) 22 *Journal of Legal Studies* 211.
[42] See, eg, P Webley *et al*, 'Studies on Minimal Rule Behaviour: Formal Rules in Public Places' (1998) 4 *Psychology Crime and Law* 309, cited in Ogus above n 25, 102.
[43] *Ibid* (Ogus) 102.

As in my example, legal intervention is an external inducement which can impose costs on actors. These costs are determined by reference to two variables: the losses incurred as a result of enforcement (such as a fixed penalty notice for parking illegally and the personal inconvenience of having to pay the fine) and the likelihood of such enforcement activity. The extent to which the law can modify behaviour is also dependent on the benefit derived from the undesirable act: in my example, the fact that parking illegally would save a considerable amount of time. In accordance with this economic framework, compliance will be achieved where the costs associated with the inducement outweigh the benefits that will be foregone if the unlawful act is not performed.

In order to evaluate the effectiveness of certain penalties in inducing behavioural change, 'it is necessary to determine and compare the cost of adapting them to the conditions of the economic model.'[44] A range of factors can influence the variables in Becker's framework. The process leading to the imposition of a sanction will, for example, influence the cost of securing that sanction. The criminal justice system, with its higher burden of proof and its traditional reliance on fault-based liability will impose significantly higher costs than say, a penalty determined by an administrative body such as a regulator. These higher costs may reduce the likelihood of non-compliant firms and individuals being pursued through the criminal courts. Another factor influencing the model relates to the costs of imposing particular sanctions so imprisonment and community service are more costly than, say, the imposition of a monetary fine or an administrative notice. Chapters 6, 7 and 8 consider in some detail, criminal, administrative and civil penalties and the procedures by which they can be imposed with a view to evaluating their effectiveness in promoting regulatory compliance.

Of course, it is important to recognise that Becker's deterrence approach does not (and cannot) reflect all the realities of regulatory enforcement.[45] However, by adopting a more expansive version of the framework, it is possible to counter many of the criticisms which have been levelled at Becker's hypothesis. As discussed in detail in Chapter 2, the expanded framework actually represents a more realistic approach to regulatory enforcement. So, for example, some critics question the very basis of the model: that individuals and firms are rational economic actors who will respond to the costs and benefits of non-compliance, notably criminal penalties. Alternative compliance and non-compliance rationales such as ignorance of the law and maintenance of corporate image may well explain regulatory behaviour and can in fact be factored into the framework by simply broadening the category of costs and benefits which are taken into account when predicting behavioural responses to enforcement. Others have difficulty in explaining how the deterrence framework can be squared with the fact that compliance rates are high when compared with the extent of enforcement. In pollution control enforcement, detection is rare, prosecution numbers are low and penalties

[44] *Ibid*, 108. See further Ogus (*Ibid*) 108–35.
[45] See further ch 2 s III.

are small.[46] But the framework can be expanded to reflect the fact that the majority of cases of non-compliance will not result in prosecution but will lead to the use of both informal enforcement responses such as warnings and formal enforcement responses other than criminal prosecution, such as monetary administrative penalties (as in a car-parking fine) and licence suspension. Both these formal and informal enforcement tools are commonly used in the context of pollution control regulation and the framework can adapted to reflect the practicalities of regulatory enforcement; that is that not all cases reach the ultimate stage of formal condemnation or indeed lead to the imposition of a formal sanction. In summary, my expanded version of the Becker framework provides an extremely useful and valuable way of analysing enforcement policy and practice. In fact, even if one remains unconvinced by the amended deterrence framework, Becker's theory can still provide some useful insights into the enforcement process.

IV A Structural Overview

In Chapters 2 and 3 of the book, I develop the theoretical framework which I will be using to evaluate enforcement policies and tools. Chapter 2 introduces Becker's framework. It considers the Becker approach as originally formulated, and examines ways in which the framework can be expanded to reflect the realities of regulatory enforcement. It is this expanded version which will form the primary tool of analysis for the remainder of the work. Chapter 3 considers the links between enforcement styles and the deterrence doctrine. The deterrence framework is based on the assumption that most regulated entities are rational economic actors whose primary aim is to maximise profits. Under this approach, a deterrence enforcement strategy, with its emphasis on both punishment and the need for mechanisms to ensure that amoral subjects find it in their best interests (in terms of cost) to comply with the law, is preferred. This is clearly at the core of my theoretical framework. But any informative and well-balanced analysis of enforcement strategies should also recognise alternative theories of enforcement such as compliance and risk-based approaches, and place them in the context of the traditional deterrence hypothesis. What is clear is that if less formalistic and more persuasive enforcement strategies are to be used, they must be underpinned by a real and credible deterrence threat. This threat, I argue, underpins enforcement strategies generally.

The remainder of the book uses the theoretical framework discussed in Chapters 2 and 3 to assess the cost-effectiveness of different enforcement tools, all of which impose different costs (whether monetary or non-monetary) on the regulated entity and regulatory authority. Examples are taken from Australia, Canada, and England and Wales. Chapter 4 sets the scene with an overview of the environmental regulatory landscape in Canada, Australia and England and Wales, focusing on

[46] See generally ch 5, and ch 7 s III D.

the institutional and legislative framework. I then move on to consider, in Chapter 5 the enforcement approach of environmental regulators in these jurisdictions.

Chapter 6 analyses the role of the criminal law in deterring recalcitrant members of the regulated community. It investigates those features of the criminal law which, under the deterrence framework, may affect its ability to deter unlawful behaviour. It considers the real versus regulatory crimes debate and the extent to which the courts' perception of regulatory crimes goes some way to explain the low level of fines imposed when sentencing polluters. It also discusses the appropriate rule of liability in the case of pollution control offences. This is clearly most relevant in the context of criminal prosecution, where a regulator may be more inclined to take criminal enforcement action when the offence is strict rather than fault-based. Whilst this may raise the likelihood of prosecution and conviction, it can potentially lower the value of the sanction imposed. Interestingly, pollution control provisions in England and Wales utilise an approach almost solely based on strict liability. In contrast, pollution control offences in Australia and Canada reflect a range of approaches from fault-based to absolute liability. A final important factor in considering the effectiveness of the deterrence model is deciding who enforcement action should be taken against. Most regulatory codes (including those in Canada, Australia and the UK), make provision for individual, corporate and director liability. Will the deterrent impact of prosecution be raised or lowered by taking action against a company director rather than the company itself?

Chapter 7 (criminal and civil sanctions) and Chapter 8 (administrative sanctions) are concerned with the 'enforcement toolbox' and the extent to which the formal, legal tools in this 'toolbox' can deter offenders and encourage compliance. In Chapter 7, I focus on the range of criminal penalties which can be imposed by the courts subsequent to conviction, although the chapter concludes with a brief examination of civil enforcement proceedings. Again, taking examples from Australia, Canada, and England and Wales, I examine the extent to which criminal sentencing options can deter individuals and corporations from committing environmental crimes, focusing on their application to the deterrence framework advanced in Chapters 2 and 3. My starting point is to consider conventional criminal sanctions such as fines and imprisonment. I then move on to analyse the effectiveness of alternative or creative sentencing options such as adverse publicity orders, environmental audit orders and environmental service orders. The courts in Australia and Canada in particular have some interesting tools at their disposal which, if used appropriately, may counter many of the difficulties of using fines to deter undesirable behaviour and may confer incidental benefits on society. In England and Wales, it would seem that the introduction of alternative sentencing options is imminent.[47] Finally, in Chapter 8, I consider the role of administrative enforcement mechanisms in promoting compliance. Administrative tools such as statutory notices and monetary administrative penalties can, in particular be extremely cost-effective as they can be served without recourse to the criminal (or indeed civil) courts.

[47] Throughout chs 6–8, reference is made to recent developments in this area in England and Wales.

2

A Theoretical Framework of Enforcement and Compliance

I Introduction

IN CHAPTER 1, I emphasised the importance of regulatory instruments (most notably command and control techniques) in government strategies designed to limit the environmental harm caused by anthropogenic activities, and recognised the need for effective enforcement if regulatory objectives are to be achieved. In this chapter, I begin my detailed analysis of the role that enforcement plays in achieving compliance with the law. The theoretical framework adopted in this, and the next chapter, forms a basis for examining pollution control law enforcement in Australia, Canada, and England and Wales. I begin section II with an analysis of the simple model of deterrence advocated by Gary Becker in his seminal paper on an economic approach to crime and punishment.[1] This assumes that individuals and firms comply with the law because of the risks and consequences of getting caught if they do not. So, in deciding whether or not to comply with the law, firms will respond to what is termed the 'expected penalty'. In other words, the model 'regards law-breaking conduct as a commodity that may be effectively "purchased" by requiring the offender to pay a penalty for breaking the law'.[2] Where the utility (or profit to the offender) of the unlawful act can be reasonably estimated, regulatory authorities will be able to predict, in theory, when an actor will be induced into complying with the law. The deterrence approach not only predicts behavioural outcomes, but also provides a theoretical basis for determining the quantum of penalties for regulatory offences.

Becker's framework is a predictive study which does not (and does not need to) consider the social desirability of the act which is the subject of enforcement action. It does not therefore make judgements about what is 'good' for society. However, in Section II C, I highlight an alternative approach to enforcement which utilises a normative theory of economics by making reference to the harmfulness of the activity. Unlike Becker's framework, this normative approach takes

[1] GS Becker, 'Crime and Punishment: An Economic Approach' (1968) 76 *Journal of Political Economy* 169.

[2] K Yeung, *Securing Compliance: A Principled Approach* (Oxford, Hart Publishing, 2004) 64.

into account the social costs of unlawful activity (including harm and enforcement costs) and, by weighing them up against the utility of the actor's behaviour, determines what constitutes socially desirable behaviour. So, where the social costs outweigh the utility to the actor, then the act is socially undesirable and should be prevented. This corresponds to the notion of optimal compliance, discussed in Chapter 1.

In utilising a law and economics approach, more specifically the deterrence hypothesis, I am not suggesting that it is the best or indeed the only framework for evaluating regulatory enforcement. Section III recognises and evaluates some of the perceived limitations of the deterrence framework. It begins in Subsection B by considering how and whether infrequent prosecutions and low sanctions (which are a common feature of regulatory enforcement) can be reconciled with the deterrence approach. Granted, one of the limitations of the deterrence hypothesis is that it was conceived by Becker with a view to determining the deterrence impact of the criminal justice system. Of course, in the regulatory enforcement context, there are a range of other enforcement mechanisms which can be used in cases of non-compliance. An expanded deterrence framework, as introduced in Subsection G, can reflect this aspect of the regulatory enforcement process. Subsections C and D explore the first assumption made under the deterrence framework: that is, that firms and individuals are motivated by wealth or utility-maximising intentions, and therefore base compliance decisions on the costs of detection and punishment and the benefits of non-compliance. As explained by Baldwin and Black:

> [C]orporate behaviour . . . is often driven not by regulatory pressure but by the culture prevailing in the sector or by the far more pressing forces of competition.[3]

Indeed, theories explaining behavioural incentives offer interesting but at times conflicting motivations for compliance (or non-compliance) and it is clear

> [T]hat no single theoretical tradition could offer a complete interpretation of all the consequences of all the total regulatory process and its effects at the company level.[4]

External inducements, namely inducements that confer benefits or impose costs on the regulated community play an important role in changing behaviour.[5] The costs and benefits approach is one of these external inducements. But external inducements extend far beyond the notion of punishment or threat of punishment. The sections also recognise the importance of what Ogus terms 'internal' inducements to comply (or not) with legal rules, such as a principled or moral obligation to follow the letter of the law.[6] However, I argue that these alternative compliance rationales do not necessarily detract from the value of the deterrence

[3] R Baldwin and J Black, 'Really Responsive Regulation' (2008) 71 *MLR* 59, 63.

[4] P Olsen, *Six Cultures of Regulation* (Copenhagen, Handelshøjskolen, 1992) 17.

[5] Inducements are categorised as being 'external' or 'internal' by AI Ogus, *Costs and Cautionary Tales—Economic Insights for the Law* (Oxford, Hart Publishing, 2006) 102 *ff.*

[6] K Hawkins, *Law As Last Resort: Prosecution Decision-Making in a Regulatory Agency* (Oxford, OUP, 2002) 259.

framework in analysing enforcement. They can be viewed, like criminal prosecution and criminal sanctions, as additional costs and benefits that must be taken into account when predicting behavioural responses. Subsections E and F briefly examine practical and principled objections to using the deterrence model as a means of predicting the deterrence impact of enforcement action.

Section III concludes by introducing a more elaborate and expanded version of Becker's deterrence framework; one that recognises important characteristics of the regulatory enforcement process and makes it more suited to our purposes. The expanded framework incorporates the use of informal sanctions (such as informal warnings), administrative penalties (such as notices and monetary administrative penalties), and civil sanctions in regulatory enforcement. It also reflects the fact that enforcement officers may progress through various 'stages' of enforcement, starting with informal enforcement tools, and moving up through the enforcement pyramid to more draconian mechanisms such as licence suspension and revocation. It is this expanded framework which is referred to in all subsequent chapters and which, I believe, facilitates a systematic analysis of the enforcement process with a view to learning important lessons about how best to achieve cost-effective enforcement.

II Inducing Behavioural Change—The Deterrence Approach

A Introduction

As discussed in Chapter 1, the aim of enforcement is to secure compliance with the law.[7] From an economic perspective, perfect compliance (and therefore perfect enforcement) is not desirable as whilst enforcement action will reduce harmful behaviour, this has to be weighed up against enforcement costs. On this basis, the aim of any enforcement agency is optimal enforcement, where the social benefits of enforcement (that is the avoided harm) equal the costs of enforcement. I, however, adopt a less ambitious mode of economic appraisal, namely enforcement which achieves a high level of compliance.

In this section, I consider enforcement not from a compliance perspective, but from a deterrence one.[8] In the criminal law context at least, deterrence or prevention is one possible theoretical justification for punishing crime.[9] Provided that its

[7] See generally ch 1 s II.

[8] A deterrence approach to punishment has long been associated with classical utilitarian philosophy advanced by proponents such as Beccaria and Bentham. See, eg, C Beccaria (1767), *On Crimes and Punishments and Other Writings* (Cambridge, CUP, 1995), and J Bentham (1789), *Introduction to the Principles and Morals of Legislation*, JL Burns and HLA Hart (eds) (Oxford, Clarendon Press, 1996).

[9] See further ch 7 s II B and the discussion of penological approaches to sentencing.

provisions are enforced with some regularity, punishment constitutes a strong disincentive to commit a crime by deterring either the convicted offender from reoffending (specific deterrence) or potential offenders from offending (general deterrence). So, from a deterrence perspective, enforcement agencies aim to prevent a particular offender from causing harm, or to deter others from causing harm. The deterrence theory of punishment has also found an important place in the economic analysis of law. Advocated by Gary Becker in 1968, the deterrence framework can be used to calculate the deterrent effects of enforcement action by viewing penalties as prices for violations.[10] In a nutshell, the model calculates the additional costs imposed on actors due to legal intervention. In accordance with the standard economic model, where these additional costs (at this stage, a criminal sanction) are exceeded by the profits accruing to the actor as a result of the offence, then the model predicts that the actor will be deterred from committing the offence, and will therefore move towards compliance. I begin this section by considering this standard economic model, referred to as 'simple deterrence' or the 'unlawful gain' model.[11] According to this model, all given acts are treated as socially undesirable (ie the expected harm is presumed to exceed the expected social benefits) and should therefore be deterred. The goal of this standard economic model is, therefore, absolute deterrence. A second policy prescription or variant of the model predicts behavioural change by balancing additional costs incurred due to legal intervention against the harm generated by the undesirable act. Unlike the simple deterrence model, this 'injury to others' model does not involve any reference to the actor's utility or profit when calculating an optimal penalty, but is based, rather, on the harmfulness of the activity.[12] As such, it performs a more ambitious function by determining the social desirability of an act. This alternative model is discussed in section C below.

B Becker's Deterrence Framework

i External Inducements v Unlawful Gains

Becker's deterrence framework is based on the positive economic justification for compliance.[13] According to this, individuals and firms are rational actors whose main objective is to maximise profits. As a result, firms will respond to positive

[10] Above n 1.

[11] Termed as such by Yeung. Above n 2, 66–8.

[12] *Ibid* 65–6.

[13] Although often grounded in economic terms, other writers, including socio-legal scholars, have recognised that additional costs to actors due to legal intervention can explain compliance behaviour. For example, Hawkins' empirical study of enforcement by Health and Safety Executive inspectors in England and Wales revealed that employers complied for 'instrumental reasons', in other words, through 'necessity (you comply because you have an economic interest in complying), or compulsion and calculation (you comply because there is a law which will penalize you if you do not and you are caught).': above n 6, 255.

and negative incentives such as subsidies and fines. Where compliance yields greater net benefits than non-compliance, legal controls will be adhered to.[14] As discussed in sections III C and D below, there are numerous other arguments put forward as to why firms and individuals comply with the law, many of which are as, if not more persuasive than the profit-maximisation approach. But this is not a reason for rejecting the valuable insights that the positive economic analysis of law can offer to a study of regulatory enforcement. As noted by Ogus:

> The fact that people do not always behave in the way envisaged in economic theorising does not by itself repudiate the theory if the assumption is plausible for the generality of cases to which it is applied.[15]

Linked to this assumption is the fact that Becker's deterrence framework works on the basis that potential offenders will respond, in particular, to incentives created by the criminal justice system. In other words, individuals and firms, in deciding whether or not to comply with the law, will base their compliance decision on what is termed the expected penalty. Where the expected penalty associated with getting caught (ie the expected cost) outweighs the expected monetary gain from committing the offence (ie the expected benefit), then the firm will be deterred from pursuing the unlawful act and will comply with the law. The expected penalty is the product of the probability of detection, prosecution and conviction (c), and the sanction (D) that would result from prosecution. The value of D includes not only the payment of the penalty, but also the cost of defending any prosecution, loss of business due to adverse publicity and possible exposure to tortious actions.[16] The deterrence approach can therefore be represented as:

$$B < cD$$

So where the benefits (B) to the actor accruing from the offence are less than the expected penalty (cD), then the actor will be induced into compliance. In other words, this is the condition for the inducement to be operative. Conversely, where the benefits (B) outweigh the expected penalty, the actor will elect to breach the regulatory controls. The framework therefore rests on the assumption that an individual will decide whether or not to commit an offence based on a personal cost/benefit calculation. So, suppose that the utility accruing from an offence (B) is £200, the probability of being detected, prosecuted and convicted (c) is 25 per cent and the sanction (D) is £1,000. Using these figures, the expected penalty (cD)

[14] Similarly, and as the counterpart of the economic theory of compliance, the positive economic justification for non-compliance sees firms and individuals as 'self-interested' or 'amoral calculators' who, in a desire to avoid the costs associated with compliance, will deliberately breach environmental obligations. See, eg Hawkins (above n 6) 230, and RA Kagan and JT Scholz, 'The 'Criminology of the Corporation' and Regulatory Enforcement Strategies' in K Hawkins and J Thomas (eds), *Enforcing Regulation* (Boston, Kluwer-Nijhoff Publishing, 1984).

[15] Above n 5, 24.

[16] CS Diver, 'A Theory of Regulatory Enforcement' (1980) 28 *Public Policy* 257, 263.

in this case would be £250 (0.25 × £1,000). Compared with the benefits of non-compliance (namely £200), the expected penalty is higher and the actor will be induced to comply.

The enforcement policy adopted by a regulatory authority can therefore affect the deterrence calculation in two ways. First, to the extent that a regulator can control the type and severity of penalty imposed (D), it is in a position to directly influence the penalty component of the expected punished. And second, a regulator can determine both the frequency with which the regulated community is monitored and inspected and the extent to which non-compliant behaviour will be prosecuted (c).[17] As illustrated in Figure I below, where the sanction (D) is a fine, the framework can calculate combinations of probability and punishment, all of which will lead to the same expected punishment. But which combination will be most cost-effective: which combination will achieve a high level of compliance at the lowest cost? If one wished to increase the level of deterrence, would it be more cost-effective to raise the expected sanction or increase the probability of detection and conviction?

The expected sanction (D) could be increased by, for example, introducing higher fines, a tactic employed by legislative bodies in Australia, Canada, and England and Wales who have raised the maximum fines available for a range of

Figure I: Expected Punishment[18]

c (probability of detection, prosecution and conviction)	*D* (sanction = fine)	*cD* (expected punishment)
1.00	10	10
0.50	20	10
0.25	40	10
0.10	100	10

pollution control offences.[19] Of course, unless judicial bodies respond by utilising the maximum range of penalties available, this in itself will not necessarily have an impact on the expected sanction. In fact, as discussed in Chapter 7 there is much evidence to suggest that the courts are not generally using the higher penalties that have been made available to them.[20] This obviously limits the deterrent impact of

[17] It is important to recognise that the basic Becker model assumes that all offenders are risk-neutral (they neither like nor dislike taking risks). Intensifying monitoring and surveillance activities would increase the objective probability of getting caught. A risk-averse offender (one who dislikes taking risks) is more easily deterred than a risk-neutral one, so the expected penalty would not need to be as high to achieve any desired degree of deterrence. Correlatively, if the offender was risk-taking, then the expected penalty will have to be higher before they are deterred from breaking the law.

[18] Taken from R Cooter and T Ulen, *Law and Economics*, 4th edn (Boston, Pearson, 2004) 474.

[19] See further ch 7 s III D.

[20] For detail see ch 7 s III D.

higher statutory maxima. There are a number of other problems with increasing fines. The most obvious of these is that the deterrence value of a fine is clearly limited by the ability of the offender to pay. If there are insufficient assets to meet the prescribed financial sanction, non-financial alternatives, such as imprisonment, have to be imposed. In the most serious cases of environmental harm, a custodial sentence may well be warranted. However, a common objection to their use in the law and economics literature is that in contrast with fines, which have a net punishment cost of zero (the defendant pays £100 and the state receives it), imprisonment imposes significant costs on society. More generally, as summarised by Friedman, 'as the punishment increases we are forced to shift to less and less efficient punishments, raising the punishment cost.'[21] One possible way of counteracting these arguments is to introduce sanctions which have indirect financial or non-financial implications such as adverse publicity orders. These are matters to which I return in Chapter 7.

Another possibility would be to raise the probability of detection (c), the second key factor in the expected penalty calculation and, as noted by the UK Environmental Audit Committee, an important driver towards compliance:

> Time and again over the course of our inquiries into environmental crime, it has been brought home to us that unless there is a real threat of being detected, the offender will continue to offend. We cannot stress strongly enough the importance of the threat of detection as a deterrent.[22]

The probability of detection does in fact rise after each apprehension as the enforcement agency becomes familiar with the offender's behavioural patterns. A raise in c could therefore be achieved by ensuring that regulated entities are monitored and inspected more frequently, thus increasing the chances of detection. This would, however, impose substantial costs on the enforcing authority and in an age of limited enforcement budgets, may not be a desirable option. Alternatively, the authority could increase the probability of prosecution (and therefore conviction). Although this too would raise enforcement costs, this approach recognises that the criminal justice system imposes a far wider range of costs than any formal sanction, and includes:

> [T]he hassle costs of pressure by an agency to comply, legal and other defence expenditures and any stigma or loss of reputation resulting from the apprehension and subsequent events.[23]

Prosecution costs could be lowered by, for example, nominating regulatory contraventions as strict liability offences. There is, however, much opposition to the use of strict liability in the criminal context and factors such as fairness may

[21] D Friedman, *Law's Order: What Economics Has to Do with Law and Why it Matters* (Princeton, Princeton University Press, 2000) 226.

[22] Environmental Audit Committee 2nd Report, *Corporate Environmental Crime* HC (8 February 2005) 136 [38].

[23] A Ogus and C Abbot, 'Pollution and Penalties' (2002) 20 *Research in Law and Economics* 493, 502.

push liability rules in a different direction.[24] Procedural changes such as lowering the burden of proof could also facilitate convictions.

ii Optimal Penalties under the Simple Deterrence Framework

Where, as is the case in the simple deterrence model, the goal of an enforcement authority is absolute deterrence, the penalty imposed on an offender should be equal to or greater than the private gain to the offender. This would eliminate any financial incentive to violate the law. The Becker approach can be used to assess the optimal sanction that should be imposed in order to deter unlawful behaviour. The optimal penalty is one with an expected value that is slightly higher than the benefit gained from non-compliance for in such a case, the penalty will create the appropriate incentive to comply with the law. So, in order to achieve the deterrence function, the sanction (D) should be greater than the utility/benefit that is derived from the unlawful act divided by the probability of detection. In other words:

$$D > \frac{B}{c}$$

So where, as before, the utility (B) is £200 and the likelihood of being detected, prosecuted and convicted (c) is 25 per cent, the optimal penalty is therefore £800 (£200/0.25).[25]

C Optimal Deterrence—The 'Injury to Others' Framework

This section would be incomplete without making reference to the normative version of Becker's simple deterrence framework. One of the main limitations of Becker's approach above is that it is predictive and is therefore based on the assumption that the relevant act is socially undesirable and should always be deterred. In other words, by imposing a penalty which at least equals the gain to the offender of committing the offence, enforcement action will always eliminate the incentive to breach the law regardless of the social desirability of the act which is the subject of enforcement. The simple deterrence approach is most appropriate where the relevant acts are assumed to be socially undesirable. However, there are some regulatory contraventions which society may not wish to deter. This is particularly so within areas of regulation, where there is usually some social utility in the activity being deterred. Also, in order for the simple deterrence framework to operate effectively, individuals and corporate entities must be willing to accept (and society must be willing to impose) large penalties for an offence that is rela-

[24] For more detail see ch 6 s II.
[25] For a discussion of possible practical objections to this, see s III E below.

tively mild in terms of the harm it causes, if the offence provides great benefits to the offender (and vice versa). Although it is conceded that this would be the exception rather than the rule, in that benefits and harm are probably usually positively correlated, it may well occur frequently enough to cause some concern.

Those charged with the responsibility of setting penalties can, however, adopt an alternative strategy which determines the quantum of penalty needed to induce the desired behaviour by reference to the amount of harm generated by the undesired act. By calculating the penalty in this way, the normative deterrence approach forces a comparison between the utility/benefit of the unlawful act and its social costs. This normative approach therefore has, as its primary aim, the maximisation of social welfare whereby socially desirable acts should be encouraged whilst socially undesirable acts should be deterred. In other words, it promotes optimal as opposed to absolute deterrence. Optimal deterrence strikes the right balance between the net cost of harm caused by the unlawful activity, and the cost of preventing it.[26] This simplified normative approach to enforcement is itself based on the presumption that firms have a straight choice between complying and not complying with the law. But it is clear that in the context of all offences, including regulatory crimes, this is not the case for most compliance decisions (obviously non-compliance may not be the result of a positive decision but may be caused by incompetence, lack of understanding etc) are ongoing; an individual may consider which of several harmful acts to commit.[27] Sanctions can also therefore play an important role in influencing which harmful acts individuals and firms choose to commit. If expected sanctions rise with the harm caused by an offence, then this will encourage individuals to commit less harmful rather than more harmful acts. Where a more harmful act is deterred because its sanction exceeds that of a less harmful act, this is commonly referred to as 'marginal deterrence'.[28]

According to Polinsky and Shavell's analysis of public law enforcement the optimal penalty (D) under this normative model can be determined by assessing the harm (H) caused by the unlawful act (including enforcement costs as part of the

[26] Economists measure optimal deterrence by comparing the social benefits of deterrence, namely reducing harmful behaviour either by deterring others from committing an offence or preventing the offender from causing harm, with the social costs of achieving a given level of deterrence. It is commonly accepted that enforcement costs tend to rise alongside increases in levels of deterrence (more enforcement officers, more prosecutors and more court time) and a point will arrive when the costs of further reductions in unlawful behaviour will not be justified by the gains. In other words, eradicating crime beyond a certain point will have a declining social benefit: 'it frequently is relatively cheap to reduce risks from 0 to 90 per cent, more expensive to go from 90% to 99% and more expensive to go from 99% to 99.9%.' (S Meyers, 'Applications of *De Minimis*' in C Whipple, *De Minimis Risk*, 9th edn (New York, Plenum Press, 1987) 102). So achieving additional reductions becomes increasingly costly. The socially optimal level of deterrence occurs at the point where the marginal social cost (that is the cost of deterring one more unlawful act) is approximately equal to the marginal social benefit arising from such crime reduction.

[27] A driver, eg, does not decide between options of speeding or not speeding—he decides how fast to drive. Similarly, in the context of industrial pollution, a factory owner may be in a position to release only a small amount of pollution into a river as opposed to a larger amount.

[28] AM Polinsky and S Shavell, 'Public Enforcement of Law' in B Bouckaert and G De Geest (eds), *Encyclopedia of Law and Economics* (Cheltenham, Edward Elgar, 2000) vol V 307, 320.

wider social costs of an unlawful activity) and the probability of detection (c).[29] The optimal penalty is that which will deter unlawful conduct to an efficient level, ie, the penalty should be set at a level equalling the total cost of the harm which it causes, thereby deterring activities which reduce social welfare. So, if we assess the value of D by reference to harm (H) rather than utility, our amended formula reads:

$$D = \frac{H}{c}$$

At its simplest, the optimal fine is the harm (H) divided by the probability of detection (c) because then, the expected fine is equal to the harm.[30] For example, let us presume that the harm is £2,000 and the probability of detection is 25 per cent. The optimal fine (H/c) is £8,000 and the expected fine is £2,000 (£8,000 × 0.25), equal to the harm caused by the unlawful act. When the expected fine equals the harm, an individual will commit the unlawful act only where the gain he derives from it (B) outweighs the harm (H) he would cause:

$$B > c \times \frac{H}{c}$$

As we have seen, this normative alternative to Becker's model of crime and punishment, by forcing a comparison between benefit and harm, actually determines when a crime is socially desirable. This has proved contentious, particularly when applied to traditional crimes such as murder and rape. Taking into account the utility of the actor is morally repugnant in relation to many mainstream crimes. However, when examining the usefulness of the hypothesis in the context of regulatory crimes such as environmental offences, many of which result from a socially beneficial activity, the formula has proved more acceptable and in many ways, is preferred over the simplified deterrence model applied previously.

D Conclusion

Whether the deterrence function of legal intervention is measured according to the benefit (utility) to the offender or the net harm caused by the unlawful act, a cost-effective system of enforcement will choose among combinations of punishment (including informal sanctions) and probability that are seen as equivalent from the point of view of the offender. The aim is to minimise the sum of punish-

[29] *Ibid* 309.

[30] The calculation becomes more complex where one considers the nature of liability and the risk approach of the offender. See further Polinsky and Shavell *Ibid* 309 *ff*.

ment and apprehension costs, thereby ensuring that the level of deterrence being aimed at is provided with the least-cost combination of punishment and probability. Environmental regulators have at their disposal a range of enforcement sanctions, formal and informal, court-based and administrative. Each sanction or penalty imposes a different set of costs on the regulator and the offender. The Becker framework is affected by these costs in two different ways. First, and most obviously, is the impact of the sanction on the offender. And second, the more expensive it is for the enforcing authority to utilise a particular sanction, the lower the probability of its imposition. These two sets of costs will also assist in determining cost-effectiveness. Chapters 6, 7 and 8 utilise this basic approach and investigate the cost-effectiveness and deterrent impact of a range of enforcement tools.

III Deterrence-based Theories—A Critical Analysis

A Introduction

In this section, I consider some of the perceived limitations of the Becker framework, which by putting a price on violations, induces regulated entities to comply with the law where the costs of non-compliance exceed the benefits to them of contravention. I begin by discussing how the deterrence approach can be reconciled with the fact that the chances of detection are slim, the taking of formal enforcement action in the event of a breach is relatively rare and where such action is taken, the type and size of penalty imposed (be it administrative, criminal or civil) is often quite low relative to the turnover of the firm.[31] I then examine alternative theories of compliance and non-compliance focusing on what can be termed 'external' and 'internal' compliance and non-compliance rationales. Non-compliance is not solely about costs and benefits and few authors would suggest that it is. Wilson, for example, states that profit maximisation is 'an incomplete statement of corporate goals.'[32] Bardach and Kagan comment that:

> Experienced regulators acknowledge that if most regulated enterprises were inclined to comply only when the threat of inspection and punishment was imminent, then the entire regulatory program would quickly collapse.[33]

These alternative rationales can, however, be incorporated (albeit imperfectly) into the deterrence approach if one is willing to expand the category of costs and benefits that are taken into account when predicting behavioural responses.

[31] See generally chs 5, 7 and 8.
[32] GK Wilson, *The Politics of Regulation* (New York, Basic Books, 1980) 359.
[33] E Bardach and RA Kagan, *Going by the Book: The Problem of Regulatory Unreasonableness* (Philadelphia, Temple University Press, 1982) 60.

Finally, I discuss the serious moral difficulties that are inherent in any deterrence-based account of enforcement and the practical problems which cast doubt on the value of the theory in guiding cost-effective enforcement.[34]

In summary, I argue that many of these limitations do not and should not detract from the important insights generated by a deterrence-based economic framework. At the end of this section, I introduce an expanded version of the deterrence approach which mirrors, more closely, the enforcement strategies and approaches of pollution control regulators. It is this expanded model which is used in subsequent analysis.

B The Deterrence Framework and Low-level Enforcement

As demonstrated in later chapters which consider pollution control law enforcement in Australia, Canada, and England and Wales, detection rates are low, prosecution is rare and penalties are small.[35] This is neither a new phenomenon, nor is it restricted to those particular jurisdictions. Empirical studies of environmental enforcement in the US reveal similar enforcement patterns, leading commentators, notably Harrington, to question how the economic theory of compliance can be reconciled with the fact that compliance rates appear to be higher than would be justified by the expected penalties approach.[36] In other words, 'if violations are rarely punished even if discovered, why would any firm bother to comply?'[37] To answer this question, referred to as the 'Harrington paradox', we need to look beyond our basic economic framework. There are several ways in which this paradox can be explained without weakening the predictive force of the deterrence approach.

Harrington himself explains the phenomenon of high compliance in the absence of strict enforcement by examining 'state-dependent' enforcement whereby the approach of the regulator will be determined by the firm's previous compliance record and status.[38] Harrington's economic model organises firms into two groups: those firms that are found to be in compliance at the last inspection (group one) and those firms that are not (group two). Where a firm is moved into group two, it will be subject to a combination of higher inspection frequency, tougher standards or higher penalties. Harrington found that a risk-neutral firm will be induced to compliance where, in the event of non-compliance, it risks being moved into group two. In other words, it is not the threat of being caught and punished per se that motivates compliance; rather it is the threat of increased monitoring, the

[34] Categorised as such by Yeung above n 2, 68–72.

[35] See generally chs 5, 7 and 8.

[36] W Harrington, 'Enforcement Leverage When Penalties are Restricted' (1988) 37 *Journal of Public Economics* 29.

[37] *Ibid* 31.

[38] *Ibid*. The model is based on a model of income tax compliance first advanced by M Landsberger and I Meilijson, 'Incentive Generating State Dependent Penalty System' (1982) 19 *Journal of Public Economics* 333.

imposition of higher standards and potentially tougher penalties.[39] This additional compliance incentive, referred to as 'enforcement leverage', goes some way to explaining compliance. It is, however, unlikely to be the sole reason.

An alternative way of reconciling the generally high levels of compliance with the relatively low expected penalties is that the informal costs (the 'hassle' costs) of detection are much higher than many have predicted. As I noted in the previous section, the costs of pressure by an agency to comply, legal and other defence costs and any loss of stigma or reputation resulting from apprehension do not depend solely on any formal sanction imposed by either the regulator or the courts. These informal costs may be a key factor in a potential offender's cost-benefit calculation. A third explanation for high compliance is that the regulated community overestimates the probability of detection and the nature of the penalties imposed in the event of a breach.[40] There is, in fact, some evidence to support this conclusion, particularly in relation to the probability of detection. Recent behavioural science literature on the ability of the criminal law to deter classic criminality suggests that citizens' perceptions of detection rates tend to be higher than the rates actually are.[41] But can the same conclusions be reached when considering regulatory infractions such as environmental crime? Empirical studies on this aspect of enforcement are rare. One study of small quantity generators of hazardous waste, conducted in the 1980s by Hammit and Reuter, found that the companies significantly overestimated the probability that an enforcement officer would detect a breach.[42] And more recently, a survey of 233 firms in eight industries in the US also revealed that 'the respondents' risk-of-detection perceptions were highly variable, ranging from close to nought per cent to 100 per cent in most industries' with the median perception of risk being relatively high, at 70 per cent.[43] However, the claim that high levels of compliance can be explained by overestimated risks of detection is weakened by the fact that the same study revealed that respondents overwhelmingly underestimated the penalty size. It would seem, therefore, that we have to look to alternative explanations for compliance.

[39] A number of regulators have adopted enforcement strategies which support Harrington's hypothesis by, eg, reducing the inspection frequency of those companies in continued compliance.

[40] For an excellent summary of the behavioural science literature on deterrence see PH Robinson and JM Darley, 'Does Criminal Law Deter? A Behavioural Science Investigation' (2004) 24 *OJLS* 173.

[41] RJ Zeckhauser and WK Viscusi, 'Risk Within Reason' in T Connolly, HR Arkes and KR Hammond (eds), *Judgment and Decision Making: An Interdisciplinary Reader*, 2nd edn (Cambridge, CUP, 2000). However, studies that focus on groups of persons who are the most likely offenders (the 'career' criminals) find that they are the more likely to realise that the probability of punishment is low and therefore perceive a lower chance of punishment than non-crime prone people. Combined with the fact that there is some evidence to suggest that many offenders tend to overestimate their own ability to make the mistakes that have led to others like them to be caught, it would seem that most criminals do not think they will be caught and punished. *Ibid* 184–5.

[42] JK Hammit and P Reuter, *Measuring and Deterring Illegal Disposal of Hazardous Waste: A Preliminary Assessment* (California, RAND, 1988).

[43] D Thornton, NA Gunningham and RA Kagan, 'Deterrence and Corporate Environmental Behaviour' 55. A paper prepared for the US EPA Corporate Environmental Behavior Research Workshop, April 2004. Available at <http://es.epa.gov/ncer/publications/workshop/pdf/EE-0484-07.pdf> accessed 5 January 2009.

C Alternative Compliance Rationales

So far, the discussion has assumed that regulated entities are profit-maximisers that respond to monitoring and sanctioning strategies when deciding whether or not to comply with the law. This is seen by many as one of the main limitations of Becker's deterrence approach in that it can only predict behavioural outcomes if potential offenders do, in fact, respond to an *ex ante* calculation that the expected benefits of non-compliance will outweigh the expected costs. In the criminal context at least, empirical research into what motivates compliance has not provided any reliable data which confirms that the cost-benefit calculation is the primary motivator of compliance. On the one hand, there is some evidence to suggest that in the case of certain offences, which tend to be committed by people who plan and think ahead, deterrence-based sentencing strategies influence the decision on whether on or not to commit the crime.[44] But on the other, there is equally persuasive evidence that points to the fact that the problem with a general deterrence-based strategy is that criminals are optimistic about the risk of being caught and ignorant about the penalty and the legal standards being enforced.[45] The cost-benefit presumption is, I believe, more persuasive in the context of social and/or economic regulation such as pollution control regulation, where the focus is on the commercial behaviour of profit-maximising firms rather than individuals, than it is when applied to traditional crime. In other words:

> [C]orporate activity is normally undertaken in order to reap some economic benefit . . . corporate decision makers choose courses of action based on a calculation of potential costs and benefits.[46]

This is clearly important as pollution offences are often committed by corporate entities rather than individuals.[47]

Although the profit-maximisation theory is arguably more persuasive when applied to social and economic regulation (including pollution control), compliance rationales undoubtedly extend beyond the expected penalty as an external inducement. It may, for example, be in the firms' self-interest to ensure compliance.[48] A good illustration of this is provided by Genn in her study of the enforcement of health and safety laws.[49] In the case of petroleum and chemical

[44] See, eg, Richard Harding's analysis of robbery. R Harding, 'Rational-Choice Gun use in Armed Robbery' (1990) 1 *Criminal Law Forum* 427.

[45] See, eg, D Riley, 'Drinking Drivers: The Limits to Deterrence' (1985) 24 *Howard Journal of Criminal Justice* 241.

[46] 'Corporate Crime: Regulating Corporate Behavior Through Criminal Sanctions' (1979) 92 *Harvard Law Review* 1227, 1235.

[47] The one exception is the waste industry which has been traditionally dominated by one-person 'cowboy' outfits.

[48] BM Hutter, *Compliance: Regulation and Environment* (Oxford, Clarendon Press, 1997) 182. Hawkins also recognises this motivation. Above n 6, 256–8

[49] H Genn, 'Business Responses to the Regulation of Health and Safety in England' (1993) 15 *Law and Policy* 219, 223.

works, where the risks of non-compliance could be so great as to involve destruction of an entire site, it was clearly in the interests of the firm to ensure compliance, thereby minimising the risk of an incident. There are, of course, less extreme examples. Complying with environmental licence conditions relating to the safe storage of solvents used in an industrial process will not only reduce the risk of environmental harm caused by a spillage but will also be economically beneficial to the company by minimising solvent wastage and thereby reducing operating costs.[50] Citing self-interest as a reason for compliance is, in many respects, more attractive than other theories. Where it is the prime motivation for compliance, it may, of course, encourage a firm to go beyond its legal obligations by, for example, reducing emissions to levels lower than strictly required under the regulations.

Another variant on the self interest theme is the desire of firms to maintain a good corporate image.[51] A major or even well-publicised minor polluting incident could, by influencing consumer and investor choice, affect the profitability of a firm. Companies perceived as environmentally irresponsible may also find themselves subject to greater scrutiny by environmental regulators and perhaps other regulatory officials. Although most companies will be concerned with their public image, Hutter found that:

> [L]arge companies were seen as particularly concerned . . . and so were those whose relationships with the general public were actually or potentially strained.[52]

With the growth of the 'green' movement and corporate social responsibility, it is suggested that the reputational costs of non-compliance will increasingly influence the behaviour of companies. These costs could of course be a direct result of prosecution. In their study of Environmental Protection Authority enforcement in the US, Badrinath and Bolster estimated that 86 per cent of the penalty for environmental prosecution in the USA is reputational.[53] Such costs could clearly also arise in the absence of prosecution and, as such, may work independently of the fact that there is a law which is enforced. As we shall see in due course, a number of jurisdictions are harnessing this reason for compliance. At the very least, informal 'naming and shaming' strategies are commonly adopted by regulatory authorities. Formal court-based publicity sanctions such as adverse publicity orders have also been introduced in a number of jurisdictions.[54]

Custom may also induce compliance. For example, Hawkins finds that, as health and safety legislation is long-established and well-recognised, 'compliance

[50] The Netregs survey (2005) reveals that 54% of SMEs saw reduced operating costs as the third most important benefit of good environmental practice (after prosecution avoidance and good customer relations). See Netregs, *SME-nvironment 2005: England* (2005) 4.

[51] Above n 48, 182. Also see AG Heyes, 'Making Things Stick: Enforcement and Compliance' (1998) 14 *Oxford Review of Economic Policy* 50, 60.

[52] Above n 48, 182.

[53] SG Badrinath, and P Bolster, 'The Role of Market Forces in EPA Enforcement' (1996) 10 *Journal of Regulatory Economics* 165.

[54] Publicity-based sanctions are discussed in detail in ch 7 s VI C.

is now regarded as the normal response to a series of well-known demands in the law.'[55] The fact that employers do not regard health and safety laws as imposing unreasonable and unexpected demands on their business, and are aware of and accept laws protecting the health and safety of employees and others, induces compliance. Bardach and Kagan, in their study of US regulation and enforcement attribute high compliance to a number of additional factors, including the threat of private law suits in response to personal injury or environmental damage resulting from regulatory non-compliance.[56] One such example is the notorious Ford Pinto case where $3.5m was paid out for deaths caused by an inadequately designed gas tank in the Pinto car. With regard to health and safety regulation, they point to both the substantial rises in workers' compensation insurance premiums following an accident and the indirect costs imposed where accident rates are high such as 'down-time, retraining expenses, high labor dissatisfaction and turnover.'[57] They also make reference to intra-organisational pressures, exerted predominantly by professionals who have some loyalty to the standards of their profession.

Internal inducements, moral convictions, or social norms (often termed principled or voluntary compliance) are also commonly cited as reasons for compliance, although without the threat of legal sanctions, these internal forces may lose much of their influence.[58] Unlike the economic model of compliance, principled compliance stems from an inner conviction that it is the right thing to do and is not the result, writes Hawkins, of 'an external compulsion' dictated by fear of sanctions.[59] Empirical studies have in fact found that internal convictions can be a significant predictor of compliance. Inspectors interviewed by Hutter and Hawkins in their (independent) studies of health and safety and the environment saw moral compliance as a strongly held sentiment among many employers. An Environment Agency study also revealed that 64 per cent of SMEs in England who are actively addressing environmental issues (or are planning to do so) are motivated by a general concern for the environment.[60] Although socially responsible firms do of course cause pollution, such occasions are likely to be perceived as accidents and the firm will normally inform the enforcing authority, willingly comply with specified clean-up and remediation requirements and take steps to prevent a repetition.[61]

There are, as I have illustrated a number of other 'external' and 'internal' factors which can and may influence behaviour. However, it would, in my view, be wrong

[55] Above n 6, 258.

[56] Above n 33, 60–66.

[57] *Ibid* 61.

[58] Many authors have recognised internal or principled compliance. For example, Hawkins above n 6, 254, Heyes above n 51, 60 and Hutter above n 48, 182–3.

[59] Above n 6, 255.

[60] Above n 50, 3.

[61] K Hawkins, *Environment and Enforcement: Regulation and the Social Definition of Pollution* (Oxford, Clarendon Press, 1984) 110–11.

to argue that the existence of these alternative compliance rationales weakens the predictive outcome of the deterrence framework. Many of these rationales can be viewed as merely another set of costs and/or benefits which should be taken into account when predicting behavioural responses. This is not to suggest that absolute values can be attributed to these external and internal factors. But they can influence the framework without the need for precise calculation. So, for example, where it is in the self-interest of a firm to comply and the firm elects to breach the regulations, then the associated consequences are merely another cost which should be factored into the predictive model. Similarly, where protecting a good corporate image motivates compliance, then the loss of reputation in the event of non-compliance is again, a cost which can be taken into account. These alternative rationales are also important in designing sanctions and penalties which will have a strong deterrence impact. Adverse publicity orders (both judicial and administrative) are just one example of a creative sentencing option which has 'tapped' in to the desire to protect corporate reputation.

D Explaining Non-compliance

Non-compliance can, of course, be explained by inter alia infrequent detection and low penalties, the absence of customary compliance norms or even difficulties in bringing civil claims for damage caused. However, there are a number of additional and potentially more persuasive rationales that can explain non-compliance with regulatory controls. One of the most commonly cited reasons for non-compliance is that firms find it difficult to comply fully with standards because they lack the technical, physical or economic capacity to do so.[62] In fact, evidence suggests that such incidents of ignorance, inability or incompetence are far more likely to explain non-compliance than the desire to minimise costs and maximise profits.[63] In some instances, companies are unable to secure compliance due to lack of awareness about the laws themselves. Recent survey evidence carried out in England would suggest that a worryingly high number of companies are ignorant about the environmental laws that apply to their process or product. Only 18 per cent of surveyed businesses in England were able to name a piece of environmental legislation unprompted.[64] Such ignorance is especially prevalent in smaller companies who have neither the resources nor the man-power to keep up-to-date with constantly changing environmental regulations.

So, ignorance etc may be viewed as a more 'genuine' reason for non-compliance than profit maximisation, although bearing in mind the possible serious and long-term consequences of environmental harm, it could be argued that operators who cannot comply with environmental standards should not be operating potentially

[62] *Ibid* 112 and above n 33, 62–4.
[63] See generally Hawkins above nn 6 and 61.
[64] Above n 50, 3.

polluting processes at all.[65] By way of illustration, take the following two contrasting examples. Company A causes a pollution incident because, due to sloppy management and careless work practices, it failed to take preventive measures recommended by the enforcing authority. Company B on the other hand demonstrates a willingness to comply but causes a pollution incident because it has neither the physical nor economic resources to implement the enforcing authority's recommendations to the full. The latter is no less an offence than the former, but the difference is that it is likely to be looked upon less harshly by a regulatory authority, culminating in a different enforcement approach.[66] Distinguishing between these two types of firms is, of course, far from easy and an inspector's perception of the reasons behind an incident may change during the course of negotiations.[67] Furthermore, a commitment to continual compliance requires companies to have good organisational and managerial capacity. Although this is often lacking, the increasing popularity of environmental management systems, even amongst small companies, has meant that this is likely to become a less frequent (and acceptable) explanation for failing to abide by legal controls.

It has also been suggested that regulated firms will not comply where they perceive the regulation and enforcement as, in some way, unreasonable. Sociological commentaries on deviant behaviour, such as 'labelling theory', suggest that levels of criminal activity are increased by the actions of lawmaking and law enforcement officials, in response for example to the prejudiced and unsympathetic policeman or the labelling of an individual as a 'delinquent'. Kagan and Scholz, in their commentary on the criminology of the corporation, apply these arguments, stating that an element of corporate non-compliance can be explained by the fact that corporations respond negatively to unreasonable laws and unreasonable enforcement. They, and others, refer to:

> [I]nstances of government arbitrariness: ill-conceived and conflicting regulations; officious and poorly trained government inspectors, unreasonable paperwork requirements; bureaucratic delay; governmental indifference to the disruption or inefficiencies in productive processes caused by literal enforcement of the law.[68]

Again, we can point to case studies which illustrate that the perceived unreasonableness of regulations increases the probability of regulatory non-compliance. For example, Kagan and Scholz were told in an interview with the corporate workplace safety director of a large company with an excellent compliance record, that he would encourage plant managers to delay regulatory compliance with laws or orders they thought were unreasonable.[69]

As discussed in the previous section, these reasons for non-compliance can again constitute a new set of costs and benefits which must be taken into account

[65] Above n 61, 112.

[66] It will also, of course, depend on the other characteristics of the offending company and its previous dealings with the enforcing authority.

[67] Above n 61, 112–13.

[68] Above n 14, 75.

[69] *Ibid.*

in predicting behavioural outcomes. They do not, in themselves, detract from the value of the deterrence framework.

E Practical Objections

It is commonly stated that attributing any real values to utility/benefit, harm and expected penalties can be extremely problematic. First and foremost, the deterrence theory can in itself generate important and valuable insights into the enforcement process without the need for precise calculation. This is, in fact, the primary aim of this monograph. This notwithstanding, presuming that those responsible for setting penalties have some knowledge of enforcement policy and judicial sentencing practice, they should be able to make a reasonable estimate of the likelihood of detection and prosecution, and the expected penalty, despite the fact that an offender's perceptions of these concepts are characterised as much by subjective judgement as by scientific precision. Putting a value on utility/benefit and harm is more difficult. Unlike many other categories of harm, environmental harm can potentially extend across boundaries, may not occur at the time of the offence (thereby increasing the possibility of future risk,) and may be difficult if not impossible to remediate. Estimating the benefits accruing to an actor as a result of non-compliance may be easier in cases where, for example, a regulated entity has avoided payment of charges or taxes. But this will not always be the case. These informational problems encountered in quantifying the value of utility or harm are partly alleviated by the fact that provided the value of the expected penalty exceeds (but is not less than) the benefit or harm, then the compliance incentive prescribed by the deterrence model will not change.[70]

F Principled Objections

There are also a number of fundamental moral or principled[71] objections to the deterrence approach, all of which can be maintained even if one were to accept that there is satisfactory empirical evidence to support a general deterrence approach. One such criticism relates to the fact that the theory could be used to support the punishment of an innocent person if to do so would successfully deter other people from committing similar offences (general deterrence).[72] In addition, Yeung points to the fact that the model 'is prone to generating counter-intuitive outcomes which may appear at odds with the community's perception of fairness and morality.'[73] For example, under the simplified deterrence model the

[70] As put by Ogus, '[I]f I am induced to obey a red light signal by a threat of £*x*, my response will not be different if the threat is £*x* plus *y*.' Above n 5, 113.

[71] Terminology used by A Ashworth, *Sentencing and Criminal Justice*, 3rd edn (London, Butterworths, 2000) 64.

[72] *Ibid* 66.

[73] Above n 2, 70.

expected cost of non-compliance (expected penalty) is determined by two interchangeable factors: the probability of detection and conviction (c) and the sanction (D). So, where the chances of being caught and convicted are extremely small (which is often the case), then the optimal penalty (whether calculated according to the utility of the offender or the harmfulness of the act) may have to be unrealistically large, and as such, may not reflect society's view of what is fair and moral. This is particularly so where the aim of an enforcement agency is absolute deterrence. Bilmes and Woodbury use the example of littering offences. Although a one-year mandatory prison sentence may, according to the model, reduce the frequency of this offence, and might be justified when one bears in mind the difficulty of detecting such infractions, 'such a penalty would be an intolerable distortion of the moral priorities of our society.'[74]

Finally, unlike desert theories of punishment, which are firmly grounded in moral notions of retribution, the deterrence theory has no moral dimension: culpability is not a factor in determining the appropriate penalty and it makes no difference if the offender acted with good or bad intentions. It is based on the idea of a rational society as opposed to a moral universe where 'the wrongness of the act may encompass something more than its costliness.'[75]

G Expanding the Simple Deterrence Framework

Many of the limitations of the deterrence framework discussed in this section can be dealt with by expanding and adapting Becker's deterrence framework. An expanded framework can reflect the fact that only a small percentage of environmental offenders are prosecuted in the criminal courts[76] and can acknowledge that where an offence is detected, a regulatory authority can use a number of different formal and informal, court-based and administrative enforcement tools.[77] Furthermore, and as I discuss in Chapter 3, an enforcement authority may adopt a responsive approach to regulation by proceeding through various stages of enforcement (of increasing severity) until it is satisfied that it has secured a particular offender's compliance. The Becker approach can be adjusted to reflect (a) the range of enforcement tools and (b) the escalation of enforcement action and can be rewritten as:[78]

[74] J Bilmes and J Woodbury, 'Deterrence and Justice: Setting Civil Penalties in the Federal Trade Commission' in RO Zerbe Jr and VP Goldberg (eds), *Research in Law and Economics Vol 14* (London, JAI Press, 1991) 211.

[75] *Ibid* 212.

[76] See generally ch 5.

[77] See further chs 7 and 8.

[78] Above n 5, 104.

$$B < gH + eF + cD$$

where *gH* represents the probability and associated costs of detection by an enforcement agency (including the hassle costs associated with detection), *eF* the probability and associated costs of administrative enforcement action (such as the imposition of a monetary administrative penalty or an enforcement notice) and *cD* the probability of formal court action (both civil and criminal) and its associated costs (including a court-based sanction). The *cD* element of the model includes, most obviously, the probability of prosecution and conviction in the criminal courts.

Figure II: A Framework of Regulatory Enforcement and Deterrence

So, using my previous analysis, in order to achieve a given level of deterrence, the enforcement agency (or where appropriate, the court) could either:

—increase monitoring and inspection frequency (g), the probability of administrative enforcement action (e) or the probability of court-based action (c); or
—increase the informal costs of detection (H), the costs of administrative enforcement action through for example serving an administrative monetary penalty (F), and the expected court-based sanction (D).

IV Conclusion

In this chapter, I have gone part way to developing a theoretical framework within which I can assess the cost-effectiveness of different enforcement mechanisms in achieving a high level of compliance with pollution control regulation. By focusing on the deterrence framework of enforcement, I have presumed that regulated firms are utility maximisers who will weigh up the costs (expected penalties) and benefits of non-compliance.[79] As neatly put by Gary Becker himself:

> The approach taken here follows the economists' usual analysis of choice and assumes that a person commits an offense if the expected utility to him exceeds the utility he could get by using his time and other resources at other activities. Some persons become 'criminals' therefore, not because their basic motivation differs from that of other persons, but because their benefits and costs differ.[80]

[79] However, as discussed in ss III C and D above, alternative compliance and non-compliance rationales can also be taken into account using Becker's framework.
[80] Above n 1, 176.

The Becker approach emphasises the role that criminal sanctions play in the compliance calculations of regulated entities and is undoubtedly a valuable way of analysing enforcement policy and practice. The model has however been criticised, in terms of both its theoretical and practical application. It presumes that the compliance decisions of firms and individuals will be influenced by a cost-benefit analysis. It also only captures 'a very small, albeit important, part of a much larger picture.'[81] It does not (and possibly cannot) catch all the complexities of the enforcement process. There are also perceived practical difficulties in establishing values such as utility and probability of detection which are important in determining the deterrence function of any given enforcement decision. Finally, there are principled objections associated with its application relating to inter alia the need for unrealistically high penalties and its lack of reference to moral aspects of crime.

I would argue, however, that some of this criticism is directed not at the model itself but rather at the role of the criminal justice system in deterring non-compliant behaviour. My expanded deterrence framework highlighted in Section III G above can counter this by recognising that regulators do not in fact rely only on criminal prosecution and criminal sanctions to deter, but use a range of other formal and informal enforcement mechanisms including informal warnings, administrative penalties and notices and civil sanctions. This arguably reflects the practical realities of modern-day regulatory enforcement.

In the chapter that follows, I complete my theoretical analysis by examining the enforcement approaches of regulators. A deterrence enforcement strategy, with its focus on prosecution and punishment is most closely linked to the simple deterrence model advanced in this chapter. Other approaches such as compliance-based strategies and 'responsive regulation' are arguably as, if not more, important in regulatory enforcement. But this should not detract from the importance of a deterrence strategy which is, I believe, of paramount importance if a more flexible enforcement approach is to be effective. It is to these matters that I shall now turn.

[81] JT Scholz, 'Enforcement Policy and Corporate Misconduct: The Changing Perspective of Deterrence Theory' (1997) 60 *Law and Contemporary Problems* 253, 254.

3

Environmental Enforcement Strategies

I Introduction

IN CHAPTER 2, I examined some of the theories that attempt to explain why and when a firm or individual will comply with its regulatory obligations, and considered how the law and economics literature on compliance and enforcement has influenced the subsequent analysis of optimal penalties. In this chapter, I consider how this literature relates to enforcement styles and strategies.

Traditionally, the regulatory enforcement debate was polarised between two seemingly opposing styles of enforcement: the 'deterrence' versus the 'compliance' approach to regulation.[1] The 'deterrence' approach, which advocates a legalistic and punitive approach to enforcement, is supported by those who believe that members of the regulated community should be regarded as 'bad apples', who contravene regulatory standards because it is cost-effective to do so. The 'compliance' approach, on the other hand, is guided towards 'good apples', who act in good faith and look to the law to guide behaviour. Here, the regulator will adopt a more softly-softly approach, using persuasion and co-operative techniques to bring firms into compliance.

However, in the past 20 to 30 years, regulatory scholars have supported much more flexible approaches to enforcement, and most would now agree that the answer to what has been termed 'the punish or persuade' debate lies in a prudent mix of the deterrence and compliance strategies. For example, the notion of 'responsive regulation' supports a tit-for-tat strategy whereby regulators will tailor their enforcement response to the particular circumstances of non-compliance.[2] This can be achieved where there is: (a) a range of enforcement responses, each of which performs a different function in the enforcement process; and (b) discretion. Of course, advances in the literature have proposed a plethora of alternative flexible strategies such as 'smart regulation', 'risk-based regulation', 'regulatory craft' and, most recently, 'really responsive regulation', none of which

[1] This should be distinguished from the deterrence hypothesis advocated by Gary Becker and discussed in detail in ch 2 s II.

[2] See generally I Ayres and J Braithwaite, *Responsive Regulation: Transcending the Deregulation Debate* (New York, OUP, 1992). Kagan and Scholz proposed a similar model: see RA Kagan and JT Scholz, 'The "Criminology of the Corporation" and Regulation Enforcement Strategies' in K Hawkins and J Thomas (eds), *Enforcing Regulation* (Boston, Kluwer-Nijhoff Publishing, 1984).

promotes the indiscriminate use of hard-hitting, deterrence-based enforcement tools.[3] As noted earlier, the economic deterrence framework advanced in Chapter 2 is most obviously aligned with the deterrence approach to enforcement. But the framework can also inform discussions on these alternative enforcement strategies such as smart or responsive regulation. This is important, as it is widely acknowledged that a purely strict deterrence approach to enforcement is undesirable. Flexible enforcement, whether it be responsive regulation, a co-operative strategy or an educational strategy, can guard against overregulation, can potentially compensate for the inherent inefficiency of uniform regulation when applied to heterogeneous situations, and may minimise monitoring and prosecution costs. Flexibility in enforcement is therefore at the core of most if not all enforcement styles. But a flexible enforcement approach depends on there being an effective and credible deterrence threat. In sum, it cannot work effectively without the presence of a deterrence threat. So the importance of deterrence should not be underestimated, even if (which is often the case), a regulatory authority adopts a much more flexible enforcement strategy.

I begin my analysis in Section II by critically analysing the deterrence and compliance approaches to enforcement. With its emphasis on utility maximisation and the role of expected sanctions in deterring criminal behaviour, the deterrence approach is most clearly aligned with the economic deterrence hypothesis advanced in Chapter 2. Compliance-based strategies, on the other hand, view persuasion and negotiation as key tools in advancing regulatory goals. My analysis of enforcement strategies continues in Section III with an evaluation of 'responsive regulation' and its associated regulatory pyramids. Responsive regulation and the regulatory pyramids were advanced as a response to the ongoing deterrence v compliance debate, and have proved to be extremely influential in the enforcement debate. I begin my assessment of responsive regulation with an overview of game theory which grounds a responsive approach to enforcement in microeconomic theory. I then consider how responsive regulation achieves a mix between deterrence and compliance-based approaches to enforcement by responding to the extent to which a regulated entity co-operates with a regulatory authority. Section III concludes by considering the benefits of discretion. Discretion is essential if a responsive (or indeed any flexible) approach to regulatory enforcement is to be adopted. In fact, discretion is a key component of enforcement even where a deterrence approach is preferred, for an enforcement officer will often have discretion in deciding between ranges of different formal enforcement options such as the initiating of a criminal prosecution or the serving of an administrative penalty or notice. I provide a brief overview of the importance of discretion (including its benefits from an economic perspective) and the mechanisms that can be used in keeping discretion in check.

Of course, responsive regulation has its limitations. It could not possibly cover all the intricacies of the enforcement process. Section IV of this chapter investi-

[3] R Baldwin and J Black, 'Really Responsive Regulation' (2008) 71 *MLR* 59.

gates concerns surrounding the conceptual underpinnings of the strategy, its application in practice and the extent to which it accords with a number of 'constitutional' values such as equal treatment, consistency and proportionality. However, despite these concerns, responsive regulation is extremely influential worldwide, and has been applied by a number of governments and regulators. It has also formed the basis for developing more complex and intricate enforcement approaches including smart regulation and really responsive regulation.[4] As such, it provides an excellent example of flexible enforcement and can be used to emphasise the importance of strong, deterrence-based sanctions in the enforcement process.

II Deterrence and Compliance Enforcement Strategies

A An Overview

The deterrence approach to enforcement is perhaps unsurprisingly most closely aligned to Becker's deterrence hypothesis, and works on the assumption that all regulated entities are profit-maximisers who assess the opportunities and risks associated with a given type of behaviour.[5] As we have seen in such cases, where the expected penalty (taking into the account the likelihood of apprehension and the likely sanction) outweighs the benefits of non-compliance, the regulated entity will pursue compliance. Furthering compliance in the case of these 'bad firms'[6] or 'amoral calculators'[7] is most suited to the use of strict, punitive monitoring and enforcement techniques. As put by Scholz:

> To implement a deterrence strategy, inspectors keep a close watch on the firm, investigate all suspicious signs, and meticulously enforce the letter of the law. Supervisors seek immediate prosecution even for trivial technical violations, and courts impose maximum penalties for wilful violations.[8]

This is not to suggest that a deterrence approach requires that all violations are prosecuted. Rather, it acknowledges that regulators may have to prosecute selectively and may, on occasion, have to bargain with these regulatees. This, however, is indicative not of co-operation but of maximising the deterrent threat that can be mustered from limited enforcement and prosecutorial resources. Using such an approach, bargaining with regulated entities is considered not only foolish but

[4] See N Gunningham and P Grabosky, *Smart Regulation* (Oxford, Clarendon Press, 1998) and Baldwin and Black, *Ibid.*
[5] See ch 2 s II.
[6] Described as such in JT Scholz, 'Co-operation, Deterrence and the Ecology of Regulatory Enforcement' (1984) 18 *Law and Society Review* 179.
[7] Above n 2 (Kagan and Scholz), 67.
[8] Above n 6, 183.

possibly also a sign of corruption. Furthermore, a regulatory officer will not make any independent judgements about the degree of seriousness of an offence, or indeed other possible reasons for non-compliance. The 'regulator as policeman' approach is largely inflexible and arguably 'encourages an unduly stark image of the world.'[9]

In contrast, the compliance enforcement approach sees actors as 'good citizens' who look to the law as a guide to proper action. Unlike the deterrence enforcement approach, non-compliance is motivated not by profit-maximisation, but by some other factor. In recognising these different motivations therefore, a compliance strategy posits that regulatory standards should, broadly speaking, be enforced in the spirit of co-operation and accommodation. As neatly summarised by Scholz:

> Like the good citizen in court, the firm is given the benefit of a doubt when wrongdoing is suspected. Regulatory inspectors come infrequently and work unobtrusively. Technical violations are overlooked if trivial, and legitimate reasons for non-compliance are accepted when warranted by circumstances. More serious violations are noted, but generous abatement periods are granted and reasonable attempts to correct the situation will forestall prosecution. When prosecution is necessary, fines are likely to be minimal, congruent with the good intentions of the firm.[10]

A compliance approach such as that described above does not negate the need for legal sanctions. Rather, the threat of legal sanctions actually enables regulators to adopt a co-operative stance; voluntary compliance would possibly be weakened if regulated entities believed that there was no threat of formal regulatory enforcement action.

The discussion so far implies a binary model of enforcement made up of compliance and deterrence approaches. This is over simplistic and enforcement strategies are arguably better represented on a sliding scale, with compliance and deterrence occupying a position at each end of the spectrum. Towards the compliance end of the spectrum, there are a range of other enforcement approaches. For example, Bridget Hutter's research into the enforcement procedures of Environmental Health Officers in Great Britain in the early 1980s identified several more specific examples of strategies which fell within the broad remit of a compliance approach: the persuasive and the insistent strategies.[11] The persuasive approach is very much like Scholz's definition of a compliance strategy cited above, with an emphasis on educating, persuading, coaxing and cajoling offenders into compliance. The insistent strategy differs in the extent to which regulators are willing to tolerate non-compliance. As stated by Hutter, regulators will adopt a persuasive stance but will, nevertheless, 'expect a fairly prompt response to their requests [and] will readily initiate legal action to effect their objectives should they encounter overt resistance to their requests.'[12]

[9] Above n 2 (Kagan and Scholz), 71.
[10] Above n 6, 183
[11] B Hutter, 'Variations in Regulatory Enforcement Styles' (1989) 11 *Law and Policy* 153.
[12] *Ibid* 156.

B The Deterrence v Compliance Debate

The compliance and deterrence approaches to enforcement arguably have the same ultimate objective, that being conformity with the law, although this would appear to be more central in a compliance approach, with deterrence strategies involving a stronger retributive dimension. The means by which they seek to promote this objective are, however, different, and for some time scholars have debated the relative merits of these two enforcement strategies.[13] Regulatory scholars are historically in the deterrence camp. Despite concluding that in some circumstances, regulatory enforcement can be too stringent, Bardach and Kagan acknowledge that a strict legalistic approach to enforcement can in fact bring about significant changes in corporate policy and management. Citing a number of examples from the United States, they found that, as a result of strict enforcement:

> [L]arger firms have scrambled to hire full-time experts . . . more in-house lawyers . . . well-trained affirmative action officers, safety engineers, toxicologists and industrial hygienists.[14]

Deterrence approaches may also reinforce society's disapproval of errant conduct and enhance social pressures to comply. Those in the 'deterrence camp' would also argue that compliance approaches to enforcement are naïve in their presumption that all regulated entities are essentially good actors and that such co-operative models are indicative of regulatory capture. This, they believe, paves the way for bad actors to take advantage of the spirit of co-operation to further their own profit-maximising goals.

In contrast, regulatory agencies tend towards the 'compliance camp'.[15] A compliance approach, they argue, involves an efficient, cost-conscious use of resources. Prosecuting through the criminal courts, or indeed using formal administrative enforcement powers, has major financial implications which limited regulatory enforcement budgets cannot stretch to. Compliance approaches, in contrast, can be justified as economically rational:

> Prosecutions are so costly in time and money that selective use of less formal mechanisms may produce higher levels of compliance for a given level of state expenditure than is possible with routine prosecution.[16]

[13] A good starting place to observe this 'battle ground' is F Pearce and S Tombs, 'Ideology, Hegemony and Empiricism: Compliance Theories of Regulation' (1990) 30 *Brit J Criminology* 423, and a response by K Hawkins, 'Compliance Strategy, Prosecution Policy and Aunt Sally: A Comment on Pearce and Tombs' (1990) 30 *Brit J Criminology* 444. The debate continued in F Pearce and S Tombs, 'Policing Corporate "Skid Rows"' (1991) 31 *Brit J Criminology* 415, and K Hawkins, 'Enforcing Regulation: More of the Same from Pearce and Tombs' (1991) 31 *Brit J Criminology* 427.

[14] E Bardach and RA Kagan, *Going by the Book: The Problem of Regulatory Unreasonableness* (Philadelphia, Temple University Press, 1982) 95.

[15] See, eg, K Hawkins, *Environment and Enforcement: Regulation and the Social Definition of Pollution* (Oxford, Clarendon Press, 1984) and P Grabosky, and J Braithwaite, *Of Manners Gentle: Enforcement Strategies of Australian Business Regulatory Agencies* (Oxford, OUP, 1986).

[16] R Baldwin and M Cave, *Understanding Regulation: Theory, Strategy and Practice* (Oxford, OUP, 1999) 98.

A good working relationship between regulator and regulated is obviously important if compliance is to be maximised. A co-operative enforcement approach is therefore much more likely to lead to information sharing and mutual respect and may also promote the adoption of more efficient methods of meeting regulatory standards. Deterrence approaches can, in contrast, promote antagonism between the two parties which as noted by Bardach and Kagan can foster 'an organised business subculture of resistance to regulation'.[17] When regulatory officials adopt a punitive approach, they compromise the willingness of the many well-intentioned, good members of the regulatory community to comply with the law. Finally, by virtue of its inherent flexibility, a compliance strategy provides regulators with a framework within which they can ignore technical violations in situations where compliance will contribute little to reducing harms.

The ongoing debate between advocates of deterrence and compliance approaches to enforcement revealed that compliance motivations differ from say firm to firm, indeed sometimes within the firm, and that adopting either a compliance or a deterrence approach would fail to reflect the real-life complexities of compliance situations. A firm may breach one legal rule due to its perceived illegitimacy but may fail to comply with a different standard because it would not be cost-effective. This paved the way for a new wave of economic and socio-legal literature on enforcement and compliance, for both regulatory and scholarly communities recognised that in order to maximise compliance with regulatory objectives, enforcing authorities should strike some kind of balance between the two models. As stated by Ayres and Braithwaite, 'The crucial question has become: When to punish; when to persuade?'[18]

In searching for an answer, the literature moves away from observational studies of behaviour which attempted to identify and understand the enforcement practices adopted by different regulatory agencies, towards more normative, methodological approaches to enforcement. One could turn to game theory, a branch of economics which studies the strategic interactions between agents, such as a firm and a regulator. One could also refer to the work of scholars such as John Braithwaite and Ian Ayres. By combining inter alia empirical observations and normative prescriptions derived from economic game theory, Ayres and Braithwaite advanced a normative model that prescribed enforcement strategies which will guide regulators in forming responses to suspected regulatory infractions. It is these matters which are discussed in section III below.

[17] Above n 14, 114.
[18] Above n 2, 21.

III Tit-For-Tat Responsive Regulation and the Enforcement Pyramid

A Game Theory and the Prisoner's Dilemma Game

A study of microeconomic theory, more specifically game theory, can assist in understanding the complexities of regulatory enforcement. Microeconomic theory is, broadly speaking, a branch of economics that is concerned with: (a) decision-making by individuals and small groups such as government authorities; and (b) how scarce resources should be allocated among competing ends. Game theory is a branch of microeconomics which can be applied to situations where there are only a few decision-makers (such as regulated and regulator) and where the optimal action of one party will depend on what the other party chooses. In a situation such as this, strategy is key, for the parties are essentially playing a game in which they respond to the actions of others. As noted by Cooter and Ulen, the 'game' is characterised by three specific things: the players, the strategies of each player, and the payoffs to each player for each strategy.[19] One of the most famous applications of game theory is the prisoner's dilemma:

> Two people, Suspect 1 and Suspect 2, conspire to commit a crime. They are apprehended by the police outside the place where the crime was committed, taken to the police station, and placed in separate rooms so that they cannot communicate. The authorities question them individually and try to play one suspect against the other. The evidence against them is circumstantial—they were simply in the wrong place at the wrong time. If the prosecutor must go to trial with only this evidence, then the suspects will have to be charged with a minor offence and given a relatively light punishment—say one year in prison. The prosecutor would very much prefer that one or both of the suspects confess to the more serious crime that they are thought to have committed. Specifically, if either suspect confesses (and thereby implicates the other) and the other does not, the non-confessor will receive seven years in prison, and, as a reward for assisting the state, the confessor will only receive half-a-year in prison. If both suspects can be induced to confess, each will spend five years in jail. What should each suspect do—confess or keep quiet?[20]

The strategy selected by Suspect 1 will determine which strategy should be selected by Suspect 2. Furthermore, each strategy has a payoff for each player. Cooter and Ulen demonstrate this by the use of a payoff matrix, which illustrates how different strategies deliver different payoffs to each player. For example, and bearing in mind that there are only two strategies—confess or keep quiet—let's say that Suspect 1 chooses to confess and Suspect 2 keeps quiet: Suspect 1 gets half-a-year in prison and Suspect 2 gets seven years. But if Suspect 1 keeps quiet and

[19] See further R Cooter and T Ulen, *Law and Economics*, 4th edn (Pearson Education, Boston, 2004) 39.

[20] *Ibid* 39.

Suspect 2 confesses, the opposite is true. Alternatively, if both suspects keep quiet, then they both get one year in prison. So, how should Suspect 1 select her optimal strategy?[21] Suspect 1 will have to go through the payoff matrix and determine the most appropriate course of action in each case, which is to confess. If Suspect 2 confesses, then Suspect 1 should confess as keeping quiet would lead to a seven-year prison sentence. If Suspect 2 keeps quiet, Suspect 1 should confess as keeping quiet would lead to a one-year prison sentence as opposed to half-a-year.

In the regulatory enforcement arena the situation is more complex as the regulated and the regulator are not playing a one-off game as in the case above, but are playing a repeated enforcement game. The work of Axelrod demonstrates that if the prisoner's dilemma game is repeated an indefinite number of times, tit-for-tat is the optimal strategy.[22] In other words, if Player 1 co-operated in the last round of the game, Player 2 should also co-operate in this round. But if Player 2 did not co-operate in the last round, then Player 1 should not co-operate in this round. Scholz, who probably made the greatest contribution to the application of game theory literature to the area of regulatory enforcement, takes Axelrod's work and uses it to advance a model of the enforcement process where the motivation of the regulated entity is to minimise regulatory costs and the motivation of the regulator is to maximise compliance outcomes.[23] He demonstrates that a tit-for-tat ('TFT') enforcement strategy will be most likely to establish mutually beneficial co-operation between the parties. As long as the firm co-operates, the regulator will adopt a co-operative approach to enforcement, but when and if the firm exploits the co-operative stance of the regulator, then the regulator will adopt a deterrent response. Similarly, as long as the regulator adopts a co-operative approach to enforcement, the firm will adopt a compliance-based strategy, but if the regulator moves to a deterrent approach, then the firm will adopt a law evasion strategy. The optimal strategy for both parties is to co-operate until the other party defects from co-operation. If and only if retaliation leads to a return to co-operation by the other party, 'the retaliator should be forgiving, restoring the benefits of mutual co-operation in place of the lower payoffs of mutual defection.'[24] In this way, the firm minimises its regulatory costs and the regulator maximises its compliance outcomes.

B Responsive Regulation

As discussed in Section II, the compliance versus deterrence debate has engaged regulatory scholars for some time, with empirical research demonstrating that a number of legal and extra-legal factors influenced the adoption of a particular enforcement strategy. In the early 1990s, the 'enforcement debate' was advanced significantly by Ayres and Braithwaite who concluded that:

21 As the game is symmetrical, the optimal strategy for Suspect 2 will be exactly the same as the optimal strategy for Suspect 1.

22 See generally R Axelrod, *The Evolution of Co-operation* (New York, Basic Books, 1984).

23 Above n 6.

24 Ayres and Braithwaite above n 2, 21.

> To reject punitive regulation is naïve; to be totally committed to it is to lead to a charge of the light brigade. The trick of successful regulation is to establish a synergy between punishment and persuasion.[25]

Responsive regulation, rather than drawing on the differences between deterrence and compliance enforcement strategies, recognises enforcement as a process involving progression through different compliance-based strategies and sanctions. Co-operation should be rewarded whilst failure to co-operate should be punished. In this way, they argued, regulation and its enforcement can be tailored to industry structure and the differing conduct of regulated firms. Whilst on the one hand, the TFT enforcement strategy discussed in the previous section is supported by Scholz because it renders co-operation the most economically rational response, responsive regulation, on the other, allows regulators to respond to both the economic and non-economic motivations of regulated entities to be law abiding.

At the core of responsive regulation is the need for a range of enforcement strategies and sanctions. These are reflected in the authors' concept of 'pyramid strategies', one of which involves a hierarchy of sanctions aimed at the individual firm (the 'enforcement pyramid'), whilst the other is concerned with regulatory strategies which relate to an entire industry (the 'pyramid of regulatory strategies'). These are represented in Figures III and IV below.

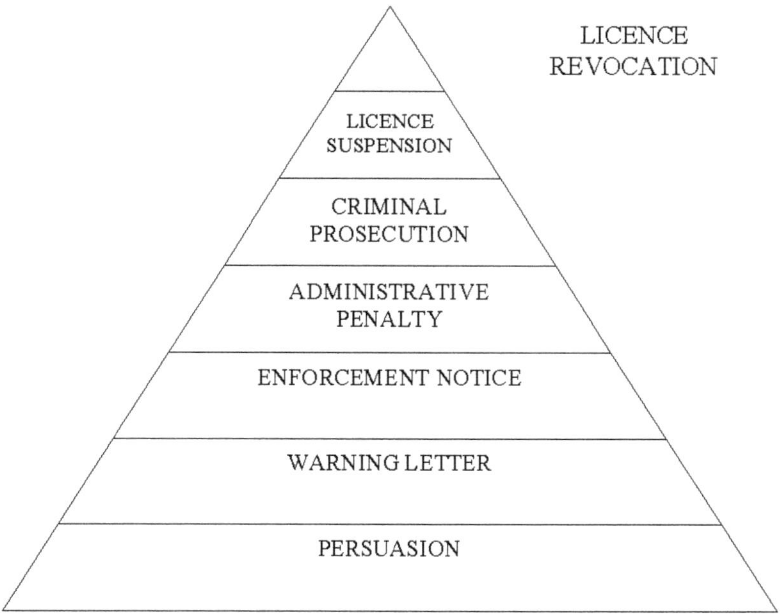

Figure III: 'Enforcement' Pyramid[26]

[25] *Ibid* 25.

[26] Based on the enforcement pyramid adopted by Ayres and Braithwaite *Ibid* 35.

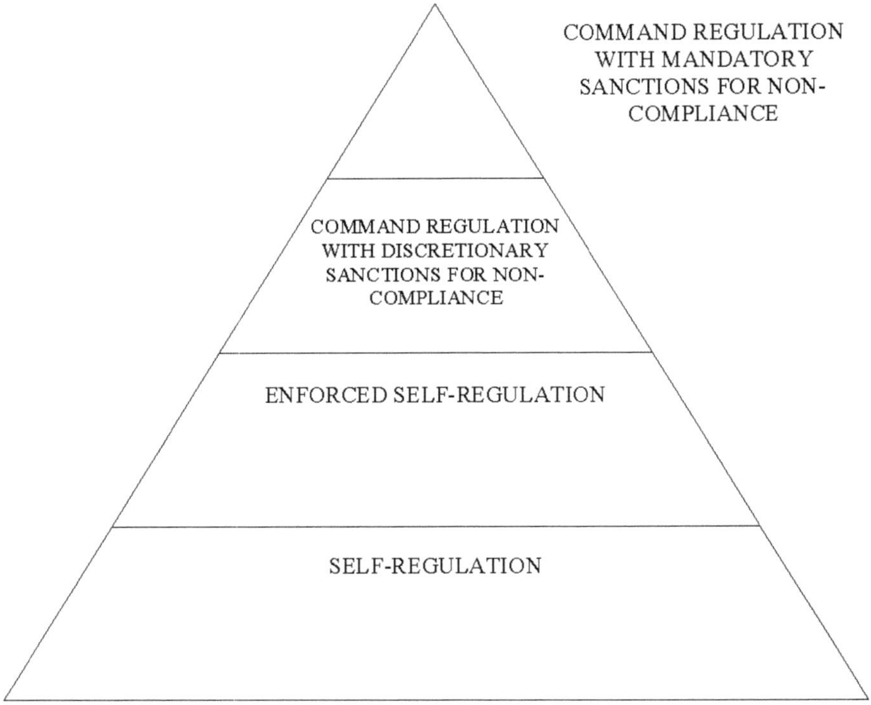

Figure IV: 'Regulatory Strategy' Pyramid[27]

In this model of responsive regulation, a firm which continues to infringe will be subjected to increasingly interventionist regulatory responses. However, where a firm comes into compliance, regulatory action will become less interventionist. The pyramid strategies reflect this by having, at their base, the least interventionist, formal, coercive and costly responses and at their apex, the most interventionist, formal, coercive and costly responses. The broad base to the pyramids reflects the fact that the majority of enforcement action will take place at this level. Similarly, the narrow apex of the pyramid, which represents the most draconian responses, is the least likely response.[28]

The content of the pyramid strategies will vary from one regulatory regime to another and different kinds of sanctioning will be appropriate to different regulatory areas. What is key in the responsive regulation model is that regulators must have recourse to a wide range of sanctions which can be applied in order of escalating severity and intrusiveness, for when regulated actors know that enforcement activity is underpinned by sanctions which can be escalated in response to varying

[27] Taken from Ayres and Braithwaite *Ibid* 39.

[28] As put by Ayres and Braithwaite themselves '[T]he key contention of this regulatory theory is that the existence of the gradients and peaks of the two enforcement pyramids channels most of the regulatory action to the base of the pyramid—in the realms of persuasion and self-regulation.' *Ibid* 39.

degrees of compliance, they are most likely to comply. The more enforcement responses the better, the theory being that:

> The taller the enforcement pyramid, the more levels of possible escalation, then the greater the pressure that can be exerted to motivate 'voluntary' compliance at the base of the pyramid.[29]

The crucial point about the pyramid is that it does not prescribe, in advance, which matters should be dealt with at its base and which at its peak. It is presumed that even in cases of serious pollution, it is better to start with persuasion at the base of the pyramid.

Yeung describes the Ayres and Braithwaite enforcement model as two-dimensional.[30] The first dimension is a static pyramid which depicts a hierarchy of sanctions graded according to their coercive backing, formality and cost. This is the 'enforcement pyramid' described above. Ayres and Braithwaite contend that regulated firms are less likely to move away from co-operation when confronted by an enforcement pyramid, than when confronted by a regulator with only one deterrence option. This is particularly so where the deterrence option is draconian and cataclysmic in the extreme. If a regulator's enforcement toolbox were limited, say, to licence suspension or revocation, its deterrent effect would be minimal, as its use would be politically impossible and morally unacceptable except in the most extraordinary of circumstances. Where this is the case, the regulator is unable to deliver a punishment payoff. Contrast this scenario with the position of a regulatory authority which has a better-equipped enforcement toolbox. In such instances, a rational actor will, in the context of a particular offence, be able to identify those tools which could practically be used and will calculate the probability that any one will be used. But the high information costs associated with determining these probabilities implies that a regulator may have superior resources with which it can 'bargain and bluff'.[31] The second 'dynamic dimension' of the enforcement pyramid refers to the TFT, responsive strategy in which the regulatory enforcement response reflects the extent to which the regulated entity has failed to co-operate. This TFT strategy will be more effective the greater number of tiers of sanctions that are available to the regulator, as illustrated by the static pyramid discussed earlier in this paragraph.

Ayres and Braithwaite also posit that regulatory enforcement will be most effective where the regulator has access to the 'benign big gun'. In their study of 96 Australian regulatory agencies, they found a cluster of regulators that adopted a co-operative enforcement strategy but that had, at the same time, some quite serious and draconian regulatory sanctions.[32] Examples included the power of the Reserve Bank to take over banks and seize gold, and the power of the Australian

[29] B Fisse and J Braithwaite, *Corporations, Crime and Accountability* (1993, Melbourne, CUP) 143, cited in F Haines, *Corporate Regulation: Beyond 'Punish or Persuade'* (Oxford, Clarendon Press, 1997) 219.

[30] K Yeung, *Securing Compliance: A Principled Approach* (Oxford, Hart Publishing, 2004) 161.

[31] *Ibid.*

[32] Above n 2 (Ayres and Braithwaite), 40.

Broadcasting Tribunal to shut down businesses completely by revoking licences. These regulators had enormous powers (including criminal prosecution) but hardly ever used them. Building this idea into the pyramid strategies, Ayres and Braithwaite conclude that the greater the punitive sanction at the apex of the pyramid, the more able is the regulator to push regulation down to the co-operative base of the pyramid.

So, in summary, Ayres and Braithwaite conclude that a co-operative approach to regulation is best predicted by:

1 the use of a tit-for-tat strategy;
2 access to a hierarchy of range of sanctions and a hierarchy of interventionism in regulatory style (the pyramid strategies); and
3 the height of the pyramid (ie the punitiveness of the most severe sanction)[33]

Of course, a key component of a flexible enforcement approach such as responsive or tit-for-tat regulatory enforcement is discretion. Discretion is arguably embedded in two key aspects of the enforcement process.[34] First, a regulatory officer has to decide whether or not regulatory standards have been complied with. In doing so, standards are treated as guidelines rather than as fixed obligations, and an officer takes into account an actor's argument in considering whether or not their behaviour is legally acceptable. Second, if a regulated entity's actions constitute 'violations', then the inspector must determine what course of action to take—whether to punish or persuade, prosecute or educate. Discretion therefore allows an officer to distinguish between serious and non-serious violations, between the basically well-intentioned regulated enterprise that can be brought into line with a warning and the recalcitrant firm that clearly deserves punishment. Prosecution can be withheld even when violations are detected and the enforcement official can act more like a 'persuader' or educator rather than a rule-bound bureaucrat. In other words, without discretion, flexibility in enforcement cannot be achieved—it is a positive way of conferring powers where it is important that officials have more freedom as to the way they are to be exercised than might be allowed by a detailed set of rules.

From a law and economics perspective, discretion is also desirable. According to the deterrence framework discussed in Chapter 2, regulated entities will respond to external inducements, most notably expected penalties, in deciding whether or not to comply with regulatory requirements. Where the probability of enforcement and the level of utility/benefit are reasonably predicable, then a fixed sanction may be economically desirable. However, in the majority of cases, it will be extremely difficult to put a value on these components, and as a result fixed sanctions will perform poorly as an external inducement.[35]

[33] *Ibid.*

[34] A definition of discretion has proved elusive. See further, eg, KC Davis, *Discretionary Justice: A Preliminary Inquiry* (Louisiana, Louisiana State University Press, 1969), R Dworkin, 'The Model of Rules' (1967) 35 *Univ of Chic LR* 14 and R Baldwin and K Hawkins, 'Discretionary Justice: Davis Reconsidered' [1984] *PL* 570.

[35] See, eg, fixed and discretionary monetary administrative penalties, discussed further in ch 8 s II.

This is not to suggest that discretion should be unstructured for unconfined discretion can lead to inconsistency, injustice and corruption. One way of confining discretion is through the imposition of rules. Rules can be used to influence substantive administrative decisions by ensuring that enforcement officials perform in accordance with official policy, particularly where those officials are implementing vague legislation.[36] Settled and precise rules are advantageous to both regulated enterprises and enforcement officials. Precise rules may limit the exposure of business to unpredictable and expensive post hoc changes and may also provide them with some assurance that they will be regulated no more stringently than their competitors. For enforcement officials, rules that specify standards in objective numerical terms facilitate proof of violations. Furthermore, rules bolster the regulatory authority's reputation for fairness, as decisions made in accordance with specific rules help rebut charges that the authority is bias. In such a climate, discretion is seen as an uncontrollable evil that leads to unpredictability, unequal treatment, a high risk of corruption, and ultimately, serious harm to unprotected citizens.[37] Alternatively, adjudicative techniques such as public inquiries and statutory tribunals can be used to approximate the formal judicial procedures that have developed in the courts.[38]

Even in the absence of these legal constraints, discretion does not remain unchecked. Practical and moral constraints have a bearing on both the substantive outcomes and the decision strategies adopted. Enforcement officials operate in a political environment; they are subjected to calls for efficient and effective operation; organisational factors can affect the way discretionary powers are exercised[39]; they are even affected by their own moral attitudes.[40] However, an appropriate balance must be achieved, for excessive use of rules and adjudicative techniques may lead to both injustice (as account cannot be taken of individual facts and circumstances) and an inefficient enforcement programme.

In sum, where a flexible approach to enforcement is preferred, enforcement officials must have discretion in applying the rules. This section has discussed one influential example of flexible enforcement, namely TFT or responsive regulation. There is no doubt that adhering to a strict deterrence or compliance enforcement strategy is unwise. Responsive regulation is just one way of mixing these two enforcement approaches in a way that promotes cost-effective enforcement and

[36] J Jowell, 'The Legal Control of Administrative Discretion' [1973] *PL* 178, 185.

[37] However, attempts to eliminate discretion and impose uniformity can never be fully successful, not least because inspectors visiting sites alone have ample opportunity to create their own discretion. Hutter claims that individual inspectors tend to have their own 'hobby horses' when inspecting premises, so different inspectors visiting the same site would pay particular attention to different aspects of waste management legislation. B Hutter, *Compliance: Regulation and Environment* (Oxford, Clarendon Press, 1997) 77–8.

[38] *Ibid* 194.

[39] For a detailed study of organisational influences on the exercise of discretion see G Richardson, A Ogus and P Burrows, *Policing Pollution—A Study of Regulation and Enforcement* (Oxford, Clarendon Press, 1982).

[40] See, eg, DJ Galligan, *Discretionary Powers—A Legal Study of Official Discretion* (Oxford, Clarendon Press, 1986) 139.

compliance. Where a regulated entity co-operates with the regulator, then enforcement action will be situated towards the base of the enforcement pyramid and persuasion, negotiation and informal enforcement tools will be used to coax the actor back into compliance. Responsive regulation is not, however, a panacea. In the next section, I consider its limitations.

IV Criticisms of Responsive Regulation

Both game theory and Ayres and Braithwaite's concept of responsive regulation (and associated enforcement pyramids) have proved valuable in prescribing the conditions under which policy-makers and regulatory authorities should adopt particular enforcement strategies in securing compliance with regulatory objectives. They not only make an important contribution to normative scholarship on regulatory enforcement but have been used by a number of governments and regulators. By 'tiering' regulatory sanctions and strategies according to the level of 'punitiveness', Ayres and Braithwaite provide a useful way of conceptualising various aspects of the enforcement process. The two-dimensional nature of the pyramid also reflects several key characteristics of enforcement including flexibility, formality, cost and timeliness, and reliance on persuasion or coercion. Furthermore, by situating such strategies and sanctions within a pyramid structure, Ayres and Braithwaite avoid depicting enforcement as a linear process:

> By envisaging enforcement tools as situated within a grade pyramid structure, it also avoids drawing a rigid dichotomy between different enforcement tools and techniques, locating them along a continuum when viewed from different functional and legal perspectives rather than being discrete and discontinuous in form and structure.[41]

However, the Ayres and Braithwaite paradigm has come up against a number of criticisms which Baldwin and Black have helpfully placed into three groups: policy or conceptual criticisms, practical criticisms and principled criticisms.[42]

A Policy/Conceptual Criticisms

There are some serious concerns about the extent to which responsive regulation is fit for purpose. One of the most commonly cited criticisms relates to the fact that in some instances of non-compliance, a step-by-step climb up the regulatory pyramid may not be appropriate. This is particularly acute in areas such as environmental protection and health and safety, where the regulatory provisions in

[41] Yeung above n 30, 162.
[42] Baldwin and Black above n 3, 62.

question are controlling potentially serious risks of harm. Where, for example, a regulatory breach exposes the possibility of pollution, it may be more apt to resort immediately to a higher level of the pyramid and indeed, as will be illustrated in Chapter 5, it is not uncommon for an environmental regulator to prosecute an offender without utilising less formal and legalistic enforcement tools first.

A second criticism relates to the fact that although moving up and down the pyramid according to the co-operation (or otherwise) of the regulated entity may be possible, moving down the pyramid may be especially difficult. This is particularly so where, due to the use of formal, punitive sanctions, what was a good relationship between the regulated entity and the regulator is compromised. Furthermore, the position and threat of 'benign big guns' at the apex of the pyramid can arguably also stand in the way of voluntary compliance, especially where compliance decisions are motivated by moral convictions or social norms that deem compliance as the 'right thing to do'.[43]

A third and important limitation of the Ayres and Braithwaite paradigm is that it presumes that the regulated community will respond to sanctions imposed by the regulator. This may not, however, be the case.[44] Where this is so, strategies can be tailored to address these alterative reasons for non-compliance. For example, non-compliance may be explained by a firm's perception of law or enforcement as being unreasonable or illegitimate. In such cases, Kagan and Scholz advance an 'inspector as politician' theory of enforcement whereby enforcement should be viewed as:

> [A] continuing effort, undertaken in particular corporate settings, amidst conflicting scientific claims and normative values, to ascertain and elicit adherence to the business procedures most likely to advance the public interest.[45]

Where strict adherence to the legal rules would impose disproportionate or unreasonable costs on a regulated firm, enforcement officers must be willing to compromise among values and persuade businesses to adopt socially acceptable solutions. Other motivations for non-compliance can generate slightly different compliance approaches. Balch, in his paper on government strategies for changing the behaviour the regulated community, advocates 'information' strategies, where non-compliance is best explained by lack of information on the regulatory requirements themselves, and 'facilitation' strategies, where a firm wants to comply but does not have the necessary time, skill or financial resources to do so.[46] Viewing a corporation as incompetent in this way suggests a 'regulatory as consultant' approach to enforcement.[47] In conclusion, these 'target-analytic' approaches[48] can

[43] *Ibid* 63. 'Principled compliance' is discussed in ch 2 s III C.

[44] For a discussion of reasons for compliance and non-compliance see ch 2 s III C and D.

[45] Above n 2 (Kagan and Scholz), 77.

[46] GI Balch, 'The Stick, the Carrot, and Other Strategies: A Theoretical Analysis of Governmental Intervention' (1980) 2 *Law and Policy Quarterly* 35, 36–7.

[47] Above n 2 (Kagan and Scholz), 80–84.

[48] Described as such by Baldwin and Black above n 3, 63.

respond to alternative motivations for non-compliance and according to Braithwaite, may in fact be more efficient than a responsive strategy.[49]

The Ayres and Braithwaite pyramids have also been criticised for making the task of regulation appear over-simplistic. For example, Haines, in her study of organisational behaviour in relation to deaths at work, recognises the utility of the pyramid structure in bringing some order to bear on the internal strategies of a single regulatory agency but, at the same time, argues that the complexities of regulatory enforcement cannot be mapped in such a simple way. Although escalation of sanction may be a perfectly reasonable response to an instance of non-compliance, Haines argues that it can, paradoxically, create distrustful organisations who will comply with minimum standards when under scrutiny but who will then retreat to non-compliance at the earliest opportunity.[50] A company's desire to avoid liability (both in statute and at common law) may also lead, in the longer term, to corporate attempts to hide internal processes from public scrutiny. If this is the case, organisational change is needed after the legal threat has subsided: a regulatory agency may not be in a position to effect such a change.

And finally, whilst Ayres and Braithwaite could not have intended that their pyramid model would capture all the complexities of the regulatory enforcement process it has nevertheless been criticised for failing to reflect the 'regulatory space' within which enforcement operates.[51] Third parties such as the insurance industry and public interest groups are, for example, key players in the regulatory process and yet are not depicted in the Ayres and Braithwaite pyramid. In other words, the pyramid is arguably of 'limited use in understanding the nature of regulation in the market-place, where multiple legal constraints and opportunities shape organizational behaviour.'[52] In response, scholars such as Grabosky have proposed a three-dimensional pyramid model with the three sides of the pyramid representing the regulator, the regulated community (including trade associations), and third parties (such as the insurance industry and public interest groups).[53] On a similar note, Haines argues that the pyramids are unable to reflect the diversity of legislation relevant to areas such as health and safety. Health and safety legislation, compensation legislation and common law liability all act with different but related purposes and most importantly, all act at the same time. Moreover, inspection and enforcement activities may be spread across different regulators. A second complexity that cannot be reflected in the pyramid paradigm is the fact that regulatory responses will have different impacts on different types of firms. Large firms may react differently to smaller firms; corporate culture

[49] Braithwaite, in his study of responsive regulation and restorative justice, proposes combining targeting and responsive approaches to regulation. J Braithwaite, *Restorative Justice and Responsive Regulation* (Oxford, OUP, 2002) 36–40. See also J Braitwaite, 'Rewards and Regulation' (2002) *Journal of Law and Society* 12, 21–2.

[50] Haines above n 29, 220.

[51] The notion of regulatory space was introduced in ch 1 s II E.

[52] Haines above n 29, 221.

[53] P Grabosky, 'Using Non-Governmental Resources to Foster Regulatory Compliance' (1995) 8 *Governance* 527.

may influence the extent to which a particular response can move a company into compliance.[54]

B Principled Objections

By adopting, as responsive regulation does, an alternative, normative approach which analyses enforcement practices with reference to a particular value or set of values, a number of principled objections are revealed. Andrew Ashworth has made an important contribution to this debate by considering the principle of equal treatment under which crimes committed in similar circumstances and of equivalent seriousness should be subjected to censure of a similar magnitude.[55] In his words,

> A system of criminal justice that allows the differential enforcement of its laws is not a system that honours the principle of equal treatment.[56]

One area of differential treatment relates to the perceived differences in enforcing 'regulatory' crimes on the one hand (with the emphasis on negotiation and co-operation) and the policing of 'ordinary' crimes on the other. Ashworth, in considering whether this can be justified, focuses on two specific responses. First, he examines the prevailing theory in Ayres and Braithwaite's responsive regulation debate, that is, that a co-operative approach in which criminal prosecution is a background threat is the most effective way of securing compliance with regulatory objectives over a long period of time. Ashworth finds this response unconvincing not least because it has yet to be demonstrated that co-operation not only generates more law-abidance but that more law-abidance can only be achieved in this way. He also examines a second possible justification for departing from the principle of equal treatment, that is that the criminal law is too blunt an instrument, and is too ineffective and costly to employ. He takes the example of fraud cases, where the cost of launching a prosecution is often given as a reason for dealing with the offence in a different way. But again, he remains unconvinced, stating that:

> If one person who wrongs another is convicted of a crime whilst another who commits an admittedly more serious wrong is not, this is manifestation of warped priorities and clear injustice.[57]

At the core of both of these responses is the view that the primary function of the criminal law is one of prevention of harm. Whilst Ashworth accepts that prevention of harm is one function of the criminal law, he argues that its main function is one of censure, a public function which should be used in as fair and

[54] Haines gives the example of adverse publicity. As we shall see in ch 7 s C the impact of adverse publicity varies according to a number of different factors.
[55] A Ashworth, 'Is the Criminal Law a Lost Cause?' (2000) 16 *LQR* 225.
[56] *Ibid* 246.
[57] *Ibid* 249.

non-discriminatory manner as possible. Equal treatment should therefore be given high priority, with any departure being justified by a legislature, court or prosecutor. In Ashworth's view, using prevention of harm as

> [A] reason for shaping the criminal law in particular ways would lead to an unacceptable distorted system in which the prospects of effectiveness and prevention, not the seriousness of the wrongdoing, would determine decisions to criminalise, decisions to prosecute and decisions about the appropriate penalty.[58]

A second source of criticism of the Ayres and Braithwaite model can be found in Karen Yeung's analysis of the enforcement of competition law. Like Haines, Yeung acknowledges the enforcement pyramid as a useful framework within which to conceptualise regulatory tools or instruments.[59] However, she identifies what she believes to be a number of serious flaws in both the static and the dynamic, prescriptive model proposed by Ayres and Braithwaite.[60] First, and closely linked to Ashworth's concerns relating to equal treatment, Yeung questions the extent to which the dynamic, TFT approach adheres to core, constitutional values such as proportionality and consistency. As has already been discussed, Ayres and Braithwaite advocate the use of a TFT strategy whereby a firm's co-operation with the regulator will be rewarded but where a firm fails to co-operate, it will be punished.[61] Yeung questions, quite legitimately in my view, whether a TFT strategy accords with the constitutional value of freedom under the law whereby individuals are free to act providing their action does not contravene the law: the TFT strategy punishes a firm that refuses to co-operate with the enforcing authority. As eloquently put by Yeung:

> There is something seriously suspicious about a strategy in which very serious regulatory infractions causing widespread harms are not treated with the appropriate level of severity because the suspect chose at all times to co-operate fully with the regulator, whilst minor infractions are dealt with in a punitive and severe manner because the suspect refused to co-operate with the regulator in its investigations.[62]

This leads Yeung to conclude that the Ayres and Braithwaite approach overlooks the constitutional values of proportionality and consistency, both of which are rooted in the right to fair and equal treatment. Like Ashworth, she supposes that a proportionate response to non-compliant behaviour must be centred on the nature and seriousness of the wrongdoing which, in turn, are informed by factors such as the degree of harm caused, the intent (or otherwise) of the wrongdoer, the duration of the contravention and the previous history of the offender. Ayres and Braithwaite, who legitimise punishment by reference to failure to persuade, therefore fail to acknowledge the importance of offence seriousness and 'neglect several

[58] *Ibid* 250.
[59] Yeung above n 30, 161–2.
[60] For a description of Yeung's static and dynamic pyramid see s B previously.
[61] See s III B.
[62] Above n 30, 168.

essential constitutional values implicated in the enforcement process.'[63] Second, Yeung argues that by portraying regulatory tools as instruments of deterrence and by focusing on the persuasion versus punishment debate, the model fails to recognise the diversity of social purposes or goals that may be advanced by these tools, such as compensation for harm suffered or restoration of damage caused.[64] Although these goals may be subsidiary or incidental to the main goal of deterrence, they are important and should be factored into the responsive regulation approach. Enforcing authorities should therefore employ the enforcement tool that will most likely achieve the desired social purpose(s).

C Practical Objections

And finally, briefly, there would also appear to be a number of practical limitations in using the pyramid approach.[65] For example, it assumes that there will be enough repeat interactions between the regulated and the regulator in order for a pyramid strategy to be operated. This may not be the case.[66] Secondly, even if there were sufficient repeat interactions to make the pyramid strategy workable, escalating through the stages of the pyramid may not happen because there are a number of other factors pulling the enforcement response in a different direction. For example, the availability of resources, the size of the regulated community and the costs of compliance may necessitate a response which does not accord with that indicated by the pyramid.[67] Furthermore, progression up the pyramid may not be politically acceptable, or indeed may not have judicial support in the way of adequate fines.[68]

V Conclusion

In this and Chapter 2, I provide a theoretical framework within which environmental enforcement strategies and sanctions can be analysed. I have taken the traditional deterrence hypothesis advocated by Gary Becker, which focuses on the role of criminal procedures and sanctions, and expanded it to encompass alternative enforcement mechanisms such as informal warnings and administrative penalties. I believe that my version of the deterrence framework depicts a more accurate representation of the complexities of modern-day regulatory enforcement in which

[63] *Ibid* 170.

[64] *Ibid* 162. For a more detailed discussion of the purposes of criminal sentencing see further Chapter 7 s II

[65] For a summary see Baldwin and Black above n 3, 64.

[66] See N Gunningham and R Johnstone, *Regulating Workplace Safety* (Oxford, OUP, 1999) 123–9.

[67] See J Mendeloff, 'Overcoming Barriers to Better Regulation' (1993) 18 *Law and Social Inquiry* 711.

[68] The problem of low fines has been felt in a number of countries including England, Canada and Australia. See further ch 7 s III D.

regulatory officials have at their disposal a range of formal and informal enforcement mechanisms.

By adopting a strictly legalistic approach to enforcement in which prosecutions are sought for even the most trivial violations, the deterrence approach, which most closely aligns with the deterrence hypothesis discussed in Chapter 2, relies on the use of formal sanctions as a means of moving regulated entities towards compliance. There is, however, limited support in the literature for the application of a purely deterrence-based enforcement style. This is reflected in enforcement practice where prosecutions are normally the exception rather than the rule, and is certainly true of pollution control law enforcement in Australia, Canada, and England and Wales.[69] But the importance of a deterrence approach should not be underestimated; flexibility in enforcement must be underpinned by a credible deterrence threat. In this chapter, responsive regulation is used as an excellent example of a flexible enforcement strategy that requires the presence of strong, deterrence-based sanctions.

A key question for us then is how, given the limited resources that enforcement agencies have to work with, the deterrence threat can be improved. According to my expanded deterrence framework, regulated entities will respond to the probability and associated costs of detection by an enforcement agency, and the probability and costs of formal conviction in the criminal (or civil) courts. Deterrence is therefore sensitive to a range of factors including inspection, detection and prosecution rates and the size and impact of the administrative penalty or, where relevant, court-based sanction. One way to enhance deterrence would be through improvements in the efficient allocation of resources by for example, shifting resources to mid-sized and larger firms who may be more aware of a regulator's general enforcement activities.[70] Another approach would be to consider techniques and procedures that could, in appropriate circumstances, be used to enhance the effectiveness of finding and penalising offenders. Scholz, for example, proposes that regulators consider techniques for improving the detection of violations such as targeting inspections, making violations more visible through better rules, and introducing information-reporting systems.[71] Alternatively, the impact of penalties could be increased by ensuring that regulators and, in the case of prosecution, the courts, have a range of penalties which can be tailored to the specific circumstances of each violation and each violator. Indirect sanctions can also be extremely powerful in deterring non-compliant behaviour. For example, the use of criminal sanctions for executives, particularly custodial sentences, can be an extremely effective way of enhancing corporate compliance due to the loss of peer approval that occurs consequent to a finding of guilt. And in the case of corporate prosecution, the monetary value of a penalty will often be insignificant compared with the potential damage to the corporate image that may accompany a conviction.

[69] See further ch 5.

[70] For a summary see JT Scholz, 'Managing Regulatory Enforcement in the United States' 427–39 in DH Rosenbloom and RD Schwartz, (New York, Marcel Dekker, 1994).

[71] *Ibid* 429–36.

I focus primarily on how legislators, regulators and the courts can enhance the impact of penalties. In subsequent Chapters, I examine inter alia rules of liability which clearly influence both the role of prosecution in the enforcement process and the setting of penalties by the courts and, where appropriate, other relevant bodies. Before doing so, in Chapters 4 and 5, I set the scene by providing an overview of the regulatory frameworks relevant to pollution control in parts of Australia and Canada, and England and Wales, and the approach of regulators to the enforcement of pollution control rules.

4

The Regulatory Landscape

I Introduction

IN THIS CHAPTER, I examine the institutional and legislative frameworks within which environmental protection goals are pursued in parts of Australia and Canada, and England and Wales. I also consider some of the pressures, economic, political and practical, faced by environmental regulators in securing compliance with pollution control regulation. This provides a basis for examining pollution control enforcement strategies in Chapter 5.

As I made clear in Chapter 1, Australia and Canada represent examples of federal systems in which the power to make law is divided between the federal government and the government of states and provinces. Each state or province has its own set of detailed pollution control regulations and penalties and its own institutional arrangements for securing compliance. So in order to make my task more manageable, I focus on pollution control regulation and its enforcement at federal level, and within one or two states or provinces. In Australia, I have chosen New South Wales (NSW) and Victoria. These are the two most populated, industrialised and therefore polluting states. NSW has a population of approximately seven million, (33 per cent of Australia's overall population) whilst Victoria is Australia's second largest state, with a population of approximately five million.[1] Both states have an active industrial and manufacturing sector[2], with emissions from industry undoubtedly being important sources of pollution. They, unsurprisingly, have well-established statutory regimes to tackle pollution. As noted by Robinson, they both have:

> [A] body of case law including decisions in the Supreme Courts and the High Court, established enforcement procedures within the EPAs, experienced legal and enforcement staff.[3]

[1] Population statistics correct as of December 2007. Australian Bureau of Statistics: <http://www.abs.gov.au> accessed 5 January 2009.

[2] Manufacturing is in fact the second largest industry in NSW. Department of State and Regional Development: <http://www.business.nsw.gov.au/industry/manufacturing> accessed 5 January 2009.

[3] B Robinson, 'The Nature of Environmental Crime' *Australian Institute of Criminology Conference Proceedings No 26* (September 1993) 3.

In Canada, I focus on enforcement in the province of Ontario. With a population of about 12 million, it is the most heavily populated province. Approximately one in three Canadians lives there and 80 per cent of Ontarians live in urban areas.[4] The province, particularly the south which is in close geographical proximity to the US market, is heavily industrialised and suffers from widespread industrial pollution.

II Australia

A Introduction

With a land mass of approximately 7.6 million square kilometres, Australia is the sixth largest country in the world. It does, however, have a relatively small (but growing) population with recent estimates at around 20.4 million.[5] In environmental terms, it is rich in biodiversity and is home to tens of thousands of species and eco-systems, a large percentage of which is unique to Australia. It unsurprisingly therefore has a wealth of 'environmental' legislation relating to inter alia natural resource management and biodiversity, marine protection and coral reef management, species and habitat protection and waste, water and air pollution.

Like many other developed nations, Australia's environmental record is somewhat mixed. Protecting Australia's biodiversity has proved to be particularly difficult and much of it's natural environment is under threat and 'in serious decline in many parts.'[6] For example, the Yearbook Australia 2007 reports that 384 species of Australian fauna are threatened; a rise of 20 per cent since 1999. 54 of these are formally classed as extinct.[7] The clearing of native vegetation to allow for enhanced agricultural production is just one of the threats to biodiversity. Fishing, shipping, oil and gas exploration and extraction, and industrial and domestic pollution also go some way to explaining why many species are in decline. Pollution is the most serious problem faced by the marine environment. Like many industrialised nations, Australia also makes a disproportionate contribution, per head of population, to global greenhouse gas (GHG) emissions, and its carbon dioxide emissions are relatively high compared with other OECD countries.[8] Driven in the

[4] Source: About Ontario: <http://www.gov.on.ca/ont/portal/!ut/p/.cmd/cs/.ce/7_0_A/.s/7_0_252/_s.7_0_A/7_0_252/_l/en?docid=004172> accessed 5 January 2009.

[5] For up-to-date population statistics see Australian Bureau of Statistics: <http://www.abs.gov.au> accessed 5 January 2009.

[6] Australia State of the Environment Committee, *Australia State of Environment 2006* ch 5 para 5.2 available at: <http://www.environment.gov.au/soe/2006/publications/report/index.html> accessed 5 January 2009.

[7] *Year Book Australia 2007*, Environment Chapter, available at <http://www.abs.gov.au/AUSSTATS/abs@.nsf/DetailsPage/1301.02007?OpenDocument> accessed 5 January 2009.

[8] *Ibid* 574.

main by economic, geographic and demographic factors, waste production is also on the increase, making Australia one of the highest waste producers in the world.[9]

As an indicator of its overall environmental performance compared with other developed countries, in 2005, the Sustainable Planning Research Group at Simon Fraser University in British Columbia, Canada published a study which compared Canada's environmental performance against that of other OECD countries.[10] The study used the most recent data verified and published by the OECD to 'rank' countries according to environmental indicators such as energy consumption, greenhouse gas emissions, pesticide use and species at risk. Taking an average across all 29 indicators, Australia was ranked joint 25th (with New Zealand) out of 30 countries, with Ireland, Canada, Belgium and the United States coming below it.

It would be wrong, however, to paint too gloomy a picture. There are many examples of how Australia has improved its environmental performance in recent years. Take waste production and recycling. Although the volume of municipal, commercial and construction waste is going up, so too the percentage of waste being recycled, with approximately 46 per cent of all waste being recycled.[11] Air quality in Australia is also improving with emissions data being made publicly available in a National Pollutant Inventory.[12] Despite some disappointing statistics, efforts to protect biodiversity have also increased substantially since 1998, with over 10 per cent of Australia's landmass now protected. Considerable efforts have also been made to reduce the environmental footprint of the agricultural sector. In fact, recent political changes in Australia indicate a potential sea change in environmental policy. The first act of the new Rudd government, elected in November 2007, was to ratify the Kyoto Protocol, leaving the United States as the only developed country not to join the Protocol. According to government sources, Australia is on track to meet its Kyoto target to reduce greenhouse gas emissions in 2008–12 to 108 per cent of its emissions in 1990. The government has also made a tougher, national commitment to reduce emissions by 60 per cent on 2000 levels by 2050. In order to meet this target, the government has inter alia set a 20 per cent renewable energy target for Australia to reach by 2020 and plans to introduce a comprehensive emissions trading scheme by 2010.

B The Institutional Framework

Environmental issues in Australia are regulated at federal, state/territorial and local level. Historically, the Commonwealth's involvement in environmental

[9] *Ibid* 568.

[10] Sustainable Planning Research Group *The Maple Leaf in the OECD: Comparing Progress Towards Sustainability 2005* (David Suzuki Foundation, 2005). Executive Summary available at <http://www.davidsuzuki.org/files/WOL/OECD-EngExec.pdf> accessed 5 January 2009.

[11] Above n 7, 569.

[12] Above n 6 ch 4 para 4.

protection issues was regarded as relatively limited, with primary responsibility lying with the state and territorial governments. However, this has arguably changed and the federal government now plays an increasingly important role in environmental matters. After examining the role of federal, state/territorial and local government in pollution control regulation, subsection 4 emphasises the tensions between federal and state environmental regulation and provides an overview of the institutions established to encourage inter-governmental co-operation.

i Federal Government

The Commonwealth or Federal Government of Australia has important legislative powers relating to the environment, with the constitutional authority to pass laws which affect the entire country.[13] Interestingly, the constitution does not contain any direct legislative powers in relation to the environment, with recent proposals to insert an 'environmental' provision via referendum not having been pursued.[14] But this has not acted as a restraint on Commonwealth powers to legislate in this area and provisions relating to trade and commerce, external affairs and taxation have all been utilised in the past to promote environmental objectives.[15] In fact, many key federal environmental statutes such as the Environment Protection and Biodiversity Conservation Act 1999 are based on more than one head of power.[16]

The Environment Protection and Biodiversity Conservation Act 1999 (EPBCA) is probably one of the most important federal statutes and is one of two legislative instruments which reflect the principles advocated in the Intergovernmental Agreement on the Environment, discussed further in subsection four below.[17] It deals with 'matters of national environmental significance' such as world heritage sites, Ramsar wetlands, endangered species, nuclear actions and the Commonwealth marine environment.[18] Broadly speaking, the Act streamlines national environmental assessment procedures. Under the Act, an approval is required for any action[19] which is likely to have a significant impact on a matter of national environmental

[13] For an overview of the role of the Federal government in environmental matters from the early 20th century to the late 1980s see M Crommelin, 'Commonwealth Involvement in Environment Policy: Past, Present and Future' (1987) 4 *Environmental and Planning Law Journal* 101.

[14] G Bates, *Environmental Law in Australia*, 5th edn (Sydney, LexisNexis, 2002) 48.

[15] For a comprehensive overview of the powers used to justify legislative enactment in the environmental sphere, see further Bates *Ibid* 55–73.

[16] This Act was based on two heads of power: external affairs and Commonwealth places.

[17] The other being the National Environment Protection Council Act 1995.

[18] 'Matters of national environmental significance' are defined in the Council of Australian Government Agreement of November 1997. For details of the Act, see generally, L Ogle, 'The EPBCA: How Workable Is It?' (2000) 17 *Environmental and Planning Law Journal* 468; CJ Campbell, 'Governance, Responsibility and the Market: Neo-liberalism and Aspects of the EPBCA' (1999) 16 *Environmental and Planning Law Journal* 290.

[19] 'Action' is defined to include a project or development, an undertaking or an activity or series of activities.

significance,[20] the environment of Commonwealth land[21] and the environment anywhere in the world (if the action is undertaken by the Commonwealth).[22] Examples of projects which may be classed as having a 'significant impact' include windfarm construction and urban, residential or commercial development. The Act contains specific enforcement and compliance mechanisms including civil and criminal liability. Prosecutions must, however, be referred to the Director of Public Prosecutions for consideration. The Department of the Environment, Water, Heritage and the Arts (discussed below) cannot bring criminal charges independently of this process. Although undoubtedly key to the protection of Australia's environment, the EPBCA and other legislation administered at federal level is not concerned primarily with pollution control regulation. This has remained within the province of states and territories.

Having outlined the legislative arm of the Commonwealth government and its role in environmental affairs, I now turn to its executive or administrative responsibilities. The Department of the Environment, Water, Heritage and the Arts (DEWHA) is currently the main government department responsible for matters of national environmental significance.[23] Its responsibilities are numerous and diverse. It advises the government on environmental policies, manages key environmental and heritage programmes and represents the government in international environmental agreements. It also administers federal environment and heritage laws including the EPBCA. The department has a total expenditure budget of approximately $1 billion.[24] Enforcement and compliance is handled by the appropriate division within the Department. For example, the EPBCA is administered by the Compliance and Enforcement Section of the Approvals and Wildlife Division. In addition to the DEWHA, there are a number of other federal government departments that clearly have environmental responsibilities. These include the Department of Climate Change, the Department of Agriculture, Fisheries and Forestry and the Department of Infrastructure, Transport, Regional Development and Local Government.

ii State/Territorial Government

In terms of regulating pollution and waste disposal, it is the six states and two self-governing territory governments that are the most important players in protecting Australia's environment.[25] In addition to setting a policy agenda, state governments

[20] Ss 11–25.
[21] Ss 26–7.
[22] S 28.
[23] Previously the Department of Environment and Water Resources.
[24] Department of Environment and Water Resources, *Annual Report 2006–2007*.
[25] The six states are Victoria, New South Wales, South Australia, Queensland, Tasmania and Western Australia. The two self-governing territories are Australian Capital Territory (ACT) and Northern Territory.

have the power to pass, administer and enforce their own laws[26] leading to eight different regulatory regimes and eight different administrative approaches to compliance and enforcement. Each state and territory has a government department charged with responsibility for environmental matters. The Department of Sustainability and Environment in Victoria is responsible for establishing policy frameworks, regulations and services to protect the environment.[27] In New South Wales, it is the Department of Environment and Climate Change.[28]

Until the mid to late 1980s, the legislative approach to pollution control regulation in most states and territories was sector specific, with separate laws on water, air, waste and noise. The main exception was Victoria which as early as 1970, passed cross-media pollution control laws in the form of the Environment Protection Act. Now, all states and territories have comprehensive, sophisticated regulatory controls which reflect key environmental principles such as pollution prevention, integrated pollution control and the taking of precautionary action. Administering and enforcing these regulations is the responsibility of a range of bodies. Many states have established a statutory environmental protection authority which takes primary responsibility for enforcing environmental laws.

One of the first states to do so was Victoria which, in the early 1970s, set up the Victoria Environmental Protection Authority (EPA (Victoria)).[29] The Authority, which was established in order to integrate environmental controls in one agency, is an independent statutory body which, like the Environment Agency in England and Wales, reports to the government (in the form of the Department of Sustainability and Environment) but at the same time is separate from government. The remit of EPA (Victoria) is extensive and covers a diversity of issues such as sustainable water management and supply, the service and management of parks and fire prevention operations.[30] It is divided into five directorates: Sustainable Development, Water and National Program, Environmental Science, Regional Services and Corporate Services. The Environment Protection Act 1970 (EPA 1970) provides the key legislative framework for regulating industry. It is described by the EPA (Victoria) as:

> [O]nly the second Act in the world to deal with the whole of the environment in a systematic and integrated way.[31]

[26] More specifically, those laws which are not controlled by the Commonwealth under s 51 of the Australian Constitution.

[27] Other important departments include the Department of Planning and Community Development which is responsible for land-use planning and environmental assessment, and the Department of Transport.

[28] Other departments with an environment portfolio include the Department of Planning and the Department of Water and Energy.

[29] For an interesting insight into environmental policy in Victoria in the 1990s see P Christoff, 'Degreening Government in the Garden State: Environment Policy under the Kennett Government, 1992–1997' (1998) 15 *Environmental and Planning Law Journal* 10.

[30] See <http://www.dse.vic.gov.au/dse/index.htm> accessed 5 January 2009.

[31] <http://www.epa.vic.gov.au/about_us/legislation/epa.asp> accessed 5 January 2009.

Furthermore, it is founded on the basic philosophy that pollution prevention and the minimisation or elimination of environmental degradation are desirable policy goals. With this purpose in mind, it sets environmental quality objectives and establishes programmes to meet them. At its centre is a licensing system whereby the operation of scheduled premises must be covered by a licence, that will set amongst other things, discharge limits and monitoring requirements. The EPA 1970 (Victoria) has evolved and been amended over the years, with the most recent amendment occurring in 2006.

In contrast, until 1992, responsibility for air, water and noise pollution in New South Wales (NSW) was split between a number of government departments and statutory authorities including the Waste Management Authority and the State Pollution Control Commission. The Environment Protection Authority (EPA NSW) was established in 1989 and charged with the responsibility of administering all the primary pollution control laws. In April 2007, the EPA was subsumed into a new Department of Environment and Climate Change (DECC) which has been structured and designed to promote an integrated approach to natural resource management. The Department is divided into eight functional areas, the most important of which for our purposes is the Environment Protection and Regulation Group:

> [It] leads the state's response to regulating a diverse range of activities that can impact on the health of the NSW environment and its people. 'Regulation' is delivered using a mix of tools, including education, partnerships, licensing and approvals, audit, enforcement and economic mechanisms.[32]

The Department has a total staff of over 3,000 and an annual expenditure of approximately $400 million. At the core of the legislative framework in NSW is the Protection of the Environment Operations Act 1997 (PEOA 1997), which sets out environmental protection policies and establishes a system of single integrated environmental licensing. The Act provides the DECC with a range of enforcement options including the serving of administrative notices and on-the-spot fines, and the initiating of criminal prosecution. Recent amendments to the Act strengthen, inter alia enforcement powers and criminal sanctions.[33]

iii Local Government

In the six states and the Northern Territory there is one further tier of government: local government. There are a number of different local authorities in each state and territory, with their powers (legislative and administrative) being defined by the relevant state/territorial government. The environmental remit of local government is, according to Bates, 'undergoing rapid transformation.'[34] Traditionally

[32] See DECC website: <http://www.environment.nsw.gov.au/whoweare/planorgs.htm> accessed 7 January 2009.

[33] Protection of the Environment Operations Amendment Act 2005.

[34] Above n 14, 84.

regarded as works authorities dealing with roads, rubbish and sewers, environmental responsibilities now extend to approval of applications for development, pollution and contaminated sites and stormwater management. Many local authorities also have enforcement powers under key pollution control legislation. In Victoria, for example, local councils can investigate and take enforcement action on certain noise issues and are responsible for issuing permits relating to septic tanks.

Local government expenditure on the environment reflects their increasingly important role. In 2002–03, local government spent over $4.1 billion on environmental management.[35] This is about 32 per cent of the total environmental expenditure of the three levels of government in Australia. Despite recent increases in local government funding for environmental matters, some argue that it could play an even more substantial role in environmental protection if it were given the appropriate incentives and legal powers to do so.[36] In response, federal government programmes have been developed to assist in providing local authorities with the skills and capacity to manage their expanding environmental remit. And legislatures have given councils new powers to eg raise environmental charges and enter into property management agreements with private landowners.

iv State/Federal Tensions and Inter-governmental Co-operation

Any analysis of environmental regulation in Australia would be incomplete without recognising the state/federal tension which underpins most Australian domestic law, and the subsequent compromises and trade-offs that this necessitates. During the 1980s, states and territories were far from accepting of the federal government's legislative activism in environmental matters. In a large number of cases, the Australian High Court was asked to rule on the constitutional validity of a number of Commonwealth Acts which were seemingly inconsistent with state laws. One such example is the Tasmanian Dam case.[37] In brief, the case concerned the Franklin Lower Gordon Wild Rivers National Park in Tasmania which had been placed on the World Heritage List by the Commonwealth government. The Australian High Court was asked to examine the validity of two pieces of Commonwealth environmental legislation, both of which were enacted under the 'external affairs' provision of the Australian constitution and were designed to prevent the Tasmanian government from building a dam on the Gordon River unless the consent of a federal Minister was granted. A Tasmanian state Act of Parliament, however, explicitly authorised the construction of the dam by the Hydro-Electric Commission and the Tasmanian government challenged the

[35] Australia State of the Environment 2006 s 11.1.

[36] For an overview, see Bates (above n 14), 85–6.

[37] *Commonwealth v Tasmania* (1983) 46 ALR 625. Also see discussion in inter alia G Bates, 'The Tasmanian Dam case and its Significance in Environmental Law' (1984) 2 *Environmental and Planning Law Journal* 325 and P Lane, 'The Federal Parliament's External Affairs Power: the Tasmanian Dam Case' (1983) 57 *Australian Law Journal* 554.

federal legislation on the grounds that it constituted a breach of the power of the States over their own land and resources. The court upheld the Commonwealth acts on the basis of the external affairs power. In this case, the Commonwealth Parliament was carrying out its international obligations under the World Heritage Convention.

During the 1980s, a number of other legal challenges served to redefine the extent of Commonwealth-state powers in favour of the Commonwealth. According to Bates, this 'upset the long held notion that legislative capacity in respect of environmental affairs rests primarily with the states.'[38] However, in the 1990s particularly from 1996 onwards when John Howard's Liberal Party was elected to office, tensions were eased. States and territories recognised the need to foster co-operative arrangements with federal government, if only to ensure that they retained some control over land and resource management issues. And the federal government acknowledged that 'the resolution of controversial resources issues by the High Court was neither appropriate nor politically acceptable.'[39]

The conclusion of an Intergovernmental Agreement on the Environment (IGAE) in 1992 heralded the beginnings of a new approach to environmental management in Australia (termed 'co-operative federalism'). In this agreement, the Commonwealth, states, territories and local government acceded to a set of environmental principles including sustainable development, integrated decision-making, cost-effectiveness and proportionality. Although the agreement recognises that states have primary responsibility for environmental management issues within their jurisdiction, it also acknowledges the legitimate and important role played by the Commonwealth government in respect of national environmental issues and established procedures which would ensure consultation and co-operation between all levels of government. Work at national level is still undertaken in the context of this agreement.

During the 1990s, national strategies and policies on the environment were also being developed[40] and ministerial councils such as the Council of Australian Governments (COAG), the Australian and New Zealand Environment and Conservation Council (ANZECC) and the National Environment Protection Council (NEPC) were established. The COAG, the principal ministerial council, deals with a number of issues including social and regulatory reform and environmental matters. Following a COAG review of Commonwealth and State involvement in environmental issues in 1997, a Heads of Agreement on Commonwealth/State Roles and Responsibilities for the Environment acknowledged the importance of the Commonwealth in environmental protection issues and inter alia specified the matters of 'national environment significance' for which the Commonwealth was responsible. In order to give effect to the agreement, existing environmental legislation was amended, leading to the Environment Protection

[38] Bates above n 14, 57. For an excellent summary of the case-law see Bates 55–73.
[39] *Ibid* 73.
[40] See eg the Ocean's Policy, the National Forests Policy and the National Wetlands Program.

and Biodiversity Conservation Act 1999 (discussed previously). The NEPC and the environmental protection components of ANZECC have since been amalgamated into the Environment Protection and Heritage Council (EPHC), the aim of which is to ensure that governments work together with communities to enhance the natural and cultural environment. One key objective of the EPHC is to agree National Environment Protection Measures (NEPMs) which establish national objectives for protecting and managing particular aspects of the environment. Current NEPMs include those relating to ambient air quality, movement of controlled waste and used packaging materials.[41]

C The Legislative Framework

Australian pollution control law dates back to the 19th century, when legislation was introduced to tackle the problems accompanying the Industrial Revolution. As in England and Wales,[42] this legislation centred not on pollution prevention and environmental protection, but on the impact of pollution on public health. It was only in the 1960s and '70s that significant and comprehensive pollution control legislation was enacted. Unlike earlier efforts, this new 'wave' of legislation did not see pollution as merely an adjunct to public health and public utilities but recognised that environmental protection was a worthwhile endeavour, in its own right. At this time, licensing systems were introduced whereby polluting entities were required to operate under a licence, and environmental and operating standards were established in order to ensure pollution minimisation or prevention. However, it was not until the 1980s that governments developed an integrated, principled and holistic approach to environmental issues. Although, in recent years, regulators have embraced 'alternative' mechanisms such as property agreements and economic instruments, it remains the case that licensing techniques play a fundamental role in protecting the environment. In most states, legislation provides a list of scheduled operations for which a licence is required. The regulatory authority charged with the responsibility of implementing and enforcing the legal requirements will then have discretion in both awarding a licence and determining the most appropriate conditions of operation. The licence applicant will normally have a merits-based appeal to a specialist court or tribunal against the decision of the regulatory authority, with much more limited provisions for third party appeals. Breach of the licensing provisions will normally constitute a criminal offence, with the threat of criminal prosecution described as 'the backbone of legal enforcement of environmental laws.'[43]

[41] For a summary see <http://www.ephc.gov.au/nepms> accessed 7 January 2009.

[42] See below s IV C.

[43] Above n 14, 204.

III Canada

A Introduction

As the world's second largest nation, covering almost 10 million square kilometres of land, Canada has an almost unprecedented wealth of natural resources. According to Environment Canada, the federal government department with primary responsibility for the environment, Canada holds around 20 per cent of the world's remaining natural areas, 25 per cent of the world's wetlands and nine per cent of the world's renewable fresh water.[44] This natural legacy is shared by only 33 million people (approximately 0.5 per cent of the world population), with urbanisation centred in a narrow band along the southern border of the country.[45]

Despite its relatively low population density, pollution and natural resource depletion have and continue to be politically significant issues. As reported in the most recently completed OECD Country Report of environmental performance in Canada in 2004, Canada has made good progress in meeting its domestic environmental objectives and international commitments.[46] But overall, Canada's environmental record is somewhat variable.[47] Building on a groundbreaking study carried out by David Boyd in 2001[48], a Maple Leaf study carried out by the Simon Fraser University and published in 2005 compared Canada's environmental performance against that of other OECD countries.[49] This study ranked Canada's overall environmental performance, across 28 environmental indicators as 28th out of 30 OECD countries, with only Belgium and the US falling below it in the rankings.[50] Air pollution would appear to be a particular problem. Although there has been a significant decrease in the emission of certain air pollutants such as sulphur dioxide, emissions of traditional air pollutants remains very high and according to the Maple Leaf study, Canada was the worst performer on two indicators relating to volatile organic compound emissions and carbon monoxide emissions. Canada was also the worst performer in relation to the generation of nuclear waste. It generated more nuclear waste *per capita* than any other OECD country and produced over six times the OECD average. On a more positive note, it had the largest decline in per capita production in nuclear waste from

[44] Environment Canada, *Tracking Key Environmental Issues* (2001) Ottawa: Minister of Public Works and Government Services 2.

[45] For up-to-date statistics on population see <http://www.statcan.ca/english/edu/clock/population.htm> accessed 5 January 2009.

[46] OECD, *Environmental Performance Reviews: Canada* (2004).

[47] See generally DR Boyd, *Unnatural Law: Rethinking Canadian Environmental Law and Policy* (Vancouver, UBC Press, 2003) ch 1.

[48] DR Boyd, *Canada vs OECD—An Environmental Comparison* (Victoria, Eco-Research Chair of Environmental Law and Policy, 2001).

[49] Above n 10.

[50] *Ibid.*

1992–2002.[51] Protection of nature and biodiversity ('ecosystem protection') has progressed and there was a 40 per cent increase in the total area of Canada under protection from 1994–2004. However, the total number of registered species at risk was increasing[52] and from 1992–2002, only two other OECD countries—Iceland and Japan—had a greater rate of increase in the number of species at risk.[53]

Law and policy has a crucial role to play in shaping preferences and influencing individual and corporate behaviour. So it is not surprising that some of the responsibility for Canada's relatively poor environmental record has been laid at the door of weak implementation and enforcement of environmental laws and regulations. As we shall see in the next section, in the second half of the 1990s implementation and enforcement efforts were seriously restricted by large cuts in federal and provincial environmental budgets. According to the OECD, although there have been subsequent increases in expenditure, 'doubts remain about the capacity to fully implement and enforce legislation and standards at federal level . . . and sometimes at provincial level.'[54]

B The Institutional Framework

In Canada, responsibility for environmental issues is divided between the federal government, the ten provinces and three territories, and local government.[55] What follows is a brief overview of the roles and responsibilities of the government bodies which operate at these three levels. The discussion will focus, at the federal level, on Environment Canada and at provincial level, the Ministry of Environment (Ontario).[56] What will become clear in this section, and in Chapter 5, is that during the 1990s, there was an 'unprecedented downsizing' of environmental departments, led by the federal government.[57] Although more money does not necessarily lead to improved enforcement, the budget cutbacks felt by environmental regulators across Canada during this time undoubtedly had an impact on enforcement activity.

[51] *Ibid*, 21–2.

[52] This is partly due to an increase in assessment work. Above n 46, 22.

[53] Above n 10, 30.

[54] Above n 46, 17. This is a matter which is discussed in detail in ch 5 s III B and C.

[55] The ten provinces are Ontario, Quebec, Nova Scotia, New Brunswick, Manitoba, British Columbia, Prince Edward Island, Saskatchewan, Alberta and Newfoundland and Labrador. The three territories are the Northwest Territories, Yukon and Nunavut.

[56] There are, of course, a number of other departments and ministries with environmental responsibilities. For example, Agriculture and Agri-Foods Canada, Foreign Affairs and International Trade Canada and Natural Resources Canada (at federal level) and the Ministry of Agriculture, Food and Rural Affairs, and the Ministry of Natural Resources (Ontario).

[57] Above n 47, 239.

i Federal Government

Environment Canada was created in 1971 and is the main government department responsible for environmental matters. Other relevant departments include Agriculture and Agri-Food Canada, Fisheries and Oceans Canada and Natural Resource Canada. Environment Canada has control over Federal crown land, (comprising approximately 40 per cent of Canada's total land area), is responsible for international and cross-border issues and administers approximately 15 federal statutes which are applicable countrywide. The Canadian Environmental Protection Act (CEPA) 1999 is the most important of these, and has, as its main objective, to make a contribution to sustainable development through pollution prevention.[58] The Act is made up of 12 parts dealing with many different aspects of environmental protection such as controlling toxic substances and managing wastes. It also contains provisions on public participation, access to information and enforcement. Another important federal act is the Fisheries Act under which it is an offence to carry on any works or undertakings that result in the harmful alteration, disruption, or destruction of fish habitat.[59] Environment Canada's Enforcement Branch has primary responsibility for enforcing the CEPA and the Fisheries Act (and corresponding regulations). Another important player in the environmental enforcement process is the Attorney-General of Canada, who also sits as the Minister in the Department of Justice. The Attorney-General has responsibility for all federal litigation and makes the final decision on whether to prosecute charges brought by Environment Canada enforcement officers.

Over the years, federal environmental enforcement activities have suffered severely from budget cuts. Over a period of 10 years from 1988 to 1998 the budget for Environment Canada fell by 30 per cent from $800 million to $550 million, more than $200 million of which was set aside for weather forecasting.[60] In terms of personnel, Environment Canada went from being the seventh largest federal department in 1988 to the smallest in 1998, at which point it had only 70 enforcement officers nationwide. According to Boyd, in 1998 there was only one enforcement officer in New Brunswick, a province with a population of 750,000, numerous pulp and paper mills and other heavy industry.[61] However, since 1998, the federal government has clearly invested in environmental enforcement activities. For example, from 1998–2001, the number of on-the-ground enforcement officers responsible for CEPA activities tripled and since 2001–02, the number of CEPA enforcement officers has risen from 92 to 105.[62]

[58] Declaration.

[59] This Act is also administered by Fisheries and Oceans Canada.

[60] G Toner, 'Environment Canada's Continuing Roller Coaster Ride' in G Swimmer (ed), *How Ottawa Spends 1996–97: Life Under the Knife* (Ottawa, Carleton University Press, 1996) 99–132.

[61] Above n 47, 240.

[62] See CEPA Annual Reports available at <http://www.ec.gc.ca/CEPARegistry/Archives/AR/Annual_reports.cfm> accessed 5 January 2009.

ii Provincial Government

It is, however, the provinces and territories that develop, apply and implement the majority of pollution control laws and policies in Canada, the jurisdiction for which is provided in the Constitutional Act 1867. Every province and territory has a ministry vested with executive powers for pollution abatement and nature conservation. Their environmental responsibilities expand across a broad array of issues including natural resource management (including water), habitat and wildlife protection and control of pollution on provincial and private land. They grant environmental permits, licences and approvals, conduct inspections and bring prosecutions for environmental offences.

The Ministry of the Environment in Ontario is responsible for:

> [P]rotecting clean and safe air, land and water to ensure healthy communities, ecological protection and sustainable development for present and future generations of Ontarians.[63]

Two of the most important pieces of environmental legislation for which the ministry is responsible are the Environmental Protection Act (EPA) 1990 and the Ontario Water Resources Act (OWRA) 1990.[64] The EPA 1990 addresses a diverse range of environmental issues from ozone depleting substances to abandoned vehicles. More generally, it prohibits the discharge of any contaminant into the natural environment in an amount, concentration or level in excess of that prescribed by the regulations.[65] The OWRA 1990, amongst other things, prohibits the discharge of sewage into the water environment and regulates the abstraction of water.

The Ministry has five divisions with their own remit and responsibilities and 26 area and district offices. The Operations Division is the operations and programme delivery arm of the Ministry and is as such, responsible for granting and administering approvals, monitoring and inspecting premises and enforcing acts and regulations. The Investigations and Enforcement Branch is an important part of the Operations Division. As suggested by its title, it is responsible for all aspects of environmental enforcement within the Ministry. In 2000, the Ministry established an Environmental SWAT Team (now the Sector Compliance Branch) to complement the inspection and enforcement work conducted by the Ministry's area and district offices. The Branch focuses on sectors that have high non-compliance rates and/or may have a significant impact on public health and the environment due to their operations. The sectors are identified using risk-assessment techniques, and unannounced inspections have been carried out in sectors such as pesticide applicators (such as farmers), electro-metal platers and industrial recycling plants. By increasing the inspection rates of specific sectors,

[63] <http://www.ene.gov.on.ca/envision/org/org-moe.htm> accessed 5 January 2009. Other important departments include the Ministry of Natural Resources and Ministry of Agriculture, Food and Rural Affairs.

[64] RSO 1990 Ch E.19 and RSO 1990 Ch O.40.

[65] S 6.

the deterrence impact of enforcement is strengthened as it raises the likelihood of non-compliant behaviour being detected. The Legal Services Branch of the Ministry is responsible for administering criminal prosecutions. However, like the system at federal level, the ultimate decision on whether to proceed with a prosecution rests with the Attorney-General.

As at federal level, the enforcement activities of the Ministry of Environment in Ontario are characterised by a series of well publicised budgetary constraints. Until 2002, Ontario was governed by the Progressive Conservative Party under Mike Harris and, from 2002, Ernie Eves. The Conservative Party promoted policies under its Common Sense Revolution platform, the key components of which were job creation, reduced government spending through deregulation and cutting 'red tape' and government restructuring.[66] There was no explicit mention of environmental protection or natural resources management and the party's subsequent Ministry of the Environment operating budget cuts reflected this stance.

As indicated in Figure V below, the operating budget of the Ministry of Environment decreased throughout most of the 1990s, hitting its lowest level in 1997–98 and 1998–99. Over a four-year period, there was a 25 per cent reduction

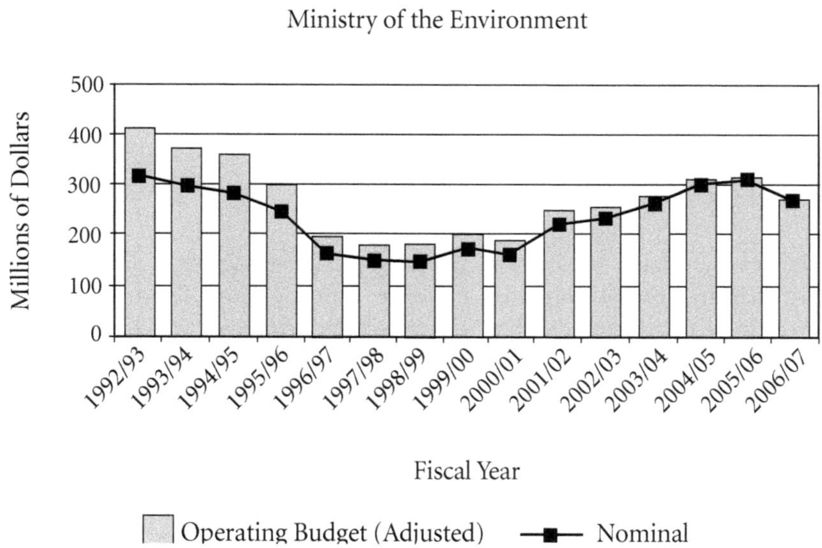

Figure V: Ministry of the Environment Operative Budget—1992–2007[67]

[66] This term was also used to describe common sense conservative platforms in New Jersey and Australia in the 1990s.

[67] Office of the Environmental Commissioner of Ontario, *Doing Less with Less: How Shortfalls in budgets, staff and in-house expertise are hampering the effectiveness of the MOE and MNR* (April 2007) 5 © Queen's Printer for Ontario, 2007. Used by permission of the ECO.

in regional staff within the Ministry.[68] As is discussed in more detail in Chapter 5, the impact of these financial cuts on inspection and enforcement activities was significant, particularly in the mid to late 1990s. According to the Canadian Institute for Environmental Law and Policy (CIELAP):

> The 1995–1999 period witnessed a precipitous decline in the province's environmental law enforcement activities. The total fines obtained by the [MOE] in 1998, the most recent year for which data could be obtained, were $863,840 - the lowest figure since 1986/1987, and less than one-third of the total for 1995. Fines fell, in part, as a consequence of the 28 per cent reduction in [IEB] staff between 1995–1998.[69]

All in all, the impact of the Common Sense Revolution is neatly summarised by Ontario's environmental commissioner who, in his 300-page analysis of Ontario's environment in 1998 stated that 'evidence of the deterioration of the province's environmental protection standards is widespread.'[70]

Since 1999, the operating budget has begun to recover. However, these increases are deceptive. First, according to a recent report by Ontario's Environment Commissioner (2007), much of that increase, until recently at least, was designed to meet drinking water protection obligations flowing from the Walkerton tragedy, and the recommendations arising from the subsequent inquiry.[71] And second, the increases do not take into account inflation or, indeed, the additional responsibilities placed on the Ministry since the early 1990s.[72] In real terms, the Ministry's operating budget has decreased from $39 per capita to $22 per capita in the 15-year period from 1992–93 to 2006–07.[73] These declines have occurred under governments formed by all three major political parties in Ontario.

iii Local Government

Under the Canadian Constitution, the provinces have, subject to limited exceptions, exclusive control over cities and other municipalities. They can, however, grant to local government, such as municipalities, any power which the province possesses. As a result, municipal government is a 'creature of the province' with the extent of its responsibilities, including those relating to environmental protection, varying according to the political will of the province. According to Benidickson,

[68] Office of the Provincial Auditor of Ontario, *Special Report on Accountability and Value for Money* (2000) ch 3.06, 119.

[69] CIELAP, *Ontario's Environment and the Common Sense Revolution—A Four Year Report* (1999) chs 1–7.

[70] Environmental Commissioner of Ontario *Open Doors: Annual Report 1998* 4.

[71] In May 2000, a deadly strain of the e.coli bacterium contaminated the drinking water system in Walkerton, Ontario. It came to light that for several days, the Public Utilities Commission confirmed the water as 'safe', despite being in receipt of laboratory test results which confirmed the outbreak. Seven people died and about 2,500 residents became ill as a result of the outbreak. For details of the inquiry see <http://www.attorneygeneral.jus.gov.on.ca/english/about/pubs/walkerton/> accessed 5 January 2009.

[72] Above n 67, 3.

[73] *Ibid* 1–2.

municipal authority on the environment 'is both vital and extensive.'[74] Reynolds has classified local authority environment-related powers into six categories: environment; public health; planning and zoning; business licensing and regulation; dangerous substances; and plenary powers.[75] It is common, however for local government to be active in issues such as the protection of public health, air and water pollution and planning and zoning. Although in recent years, the courts have been willing to acknowledge the importance of the role of local government in environmental issues, there is widespread concern at the lack of financial resources available to municipalities to pursue desirable initiatives.[76]

iv Inter-governmental Co-operation and Consultation

With environmental responsibilities being spread across a three-tier system of government, it is essential to ensure, as in Australia, that approaches to inter-jurisdictional issues such as air pollution and toxic chemicals are co-ordinated and mechanisms are in place to facilitate adequate and effective co-operation between jurisdictions. In this respect, the most important institution dealing with environmental matters is the Canadian Council of Ministers of the Environment (CCME). The Council is made up of 14 environmental ministers from the federal, provincial and territorial governments and although it has no power to implement and enforce legislation and is therefore unable to impose its standards on governments, its members can collectively establish national environmental standards, strategies and objectives. Over the past few years, the CCME has focused on initiatives which work towards the objectives of the Canada-wide Accord on Environmental Harmonization, which was signed in 1998.[77] Under the Accord, governments are called upon 'to work in partnership to reach the highest possible level of environmental protection for all Canadians.'[78] To assist in achieving this vision, the Accord sets out the objectives of harmonisation and the principles that should be reflected in the government's environmental management activities.[79] In compliance with the Accord, governments, through the CCME, have implemented a number of sub-agreements on Canada-wide standards and environmental assessment. For example, ambient air quality standards for ground-level ozone have been set and there is a national objective to reduce benzene.

[74] J Benidickson, *Environmental Law*, 2nd edn (Toronto, Irwin Law Inc, 2002) 34.

[75] LA Reynolds, 'Environmental Regulation and Management by Local Public Authorities in Canada' (1993) 3 *Journal of Environmental Law and Practice* 41, cited in Benidickson *Ibid* 34.

[76] For more information see Benidickson *Ibid* 34–6.

[77] The Accord and its sub-agreements have not been ratified by Quebec.

[78] As stated in the Accord's 'vision'.

[79] The objectives include enhanced environmental protection and the promotion of sustainable development. The Accord's principles include the polluter pays principle and the precautionary principle.

C The Legislative Framework

Like many other industrialised nations, Canadian statutes dealing exclusively with the discharge of waste sources into the environment developed in the 1970s. These basic air, water and land pollution provisions adopted a 'command and control' approach to regulation. Permits were issued by a state regulator and terms and conditions attached to those permits controlled the quality and quantity of any waste discharge.[80] Failure to comply with the regulatory requirements was a criminal offence, punishable on summary conviction by modest fines. Many of the statutes also included a more general offence of causing harm to human life or health or the environment. Since the 1970s, most of the statutes have been amended and fine-tuned by policies, regulations and procedures. Webb notes, during the 1980s, a discernable shift away from absolute prohibitions to the introduction of more interventionist restrictions based on acceptable levels of pollutants, the practicability of abatement techniques, the cost to industry and public opinion.[81]

Lucas also identifies the emergence of a 'second generation' of environmental statutes, the main objective of which is to control persistent toxic substances which pose a threat to the environment. By their nature,

> [S]uch small doses of toxicity and slow decomposition characteristics make the established assimilative waste regulation approach unsuitable for dealing effectively with toxic substances.[82]

In addition, these new laws recognise the intergenerational effects of environmental damage and acknowledge that the law must be flexible and able to respond to new scientific and technical knowledge about the toxicity of certain dangerous substances. They also recognise the transboundary nature of environmental harms: provincial laws are designed to respect Canada's international commitments as well as inter-provincial and federal-provincial targets and understandings whilst federal laws clearly implement specific international environmental commitments. Another key feature of these 'new' environmental statutes is the presence of more sophisticated enforcement provisions relating to inter alia the issuing of tickets for minor contraventions, the availability of administrative orders and the introduction of serious indictable criminal offences for actions that endanger life or health.

Licences, permits or certificates of approval still play a pivotal role in controlling potentially harmful activities and, depending on the nature and scope of a

[80] The Ontario Environmental Protection Act and the Quebec Environmental Quality Act are good examples of the traditional 'command and control' approach.

[81] K Webb, 'Pollution Control in Canada: The Regulatory Approach in the 1980s' (1988) *Study Paper, Administrative Law Series* (Ottawa, Law Reform Commission of Canada) 14–15.

[82] AR Lucas, 'The New Environmental Law' in RL Watts and DM Brown (eds), *Canada: The State of the Federation: 1989* (Kingston, Queen's University, Institute of Intergovernmental Affairs, 1990) 168–171 cited in EL Hughes, AR Lucas and WA Tilleman (eds), *Environmental Law and Policy*, 2nd edn (Toronto, Emond Montgomery Publications, 1998) 141.

proposed activity, such approvals are required as a precondition to development or operations. The scope of terms and conditions attached to these certificates implement general (and more specific) statutory environmental standards although much is left to the discretion of the decision-maker.[83] Environmental statutes also create a range of offences. The precise details of such offences, at provincial and federal level vary significantly but broadly speaking, they follow three different 'models'. Many statutes create a general offence of discharging a contaminant which is likely to cause harm to the environment.[84] It is also often an offence to discharge in excess of standards prescribed in regulations.[85] Rather than prohibit pollution *per se*, many offences ensure compliance with the administrative framework.[86]

IV England and Wales

A Introduction

The UK is a small but relatively highly populated country with an area of approximately 242,000 square kilometres and according, to the Office of National Statistics, a population of about 59.6 million people in June 2003.[87] England and Wales has an area of approximately 133,000 square kilometres and a population of approximately 53 million.[88] Approximately 84 per cent of the UK's population lives in England and 5 per cent in Wales.[89] Compared with Canada and Australia which have population densities of 3.2 and 2.6 people per square kilometre respectively, the UK's figure of 246 people per square kilometre makes it the 51st most densely populated country in the world.

The UK's large (and growing) population is one factor which clearly puts serious pressures on the environment and natural resources. The UK's position as the cradle of the Industrial Revolution is undoubtedly also responsible for much of the country's historical pollution. However, with industry's contribution to GDP dropping in recent years, the OECD has noted that:

[83] See, eg, Environmental Protection Act (Ontario) s 9(4) and (5).

[84] See, eg, Environment Quality Act (Quebec) s 20, Water Resources Act (Ontario) s 30(1) and Environmental Protection Act (Ontario) s 14.

[85] Environmental Protection Act (Ontario) s 6 and Environmental Protection and Enhancement Act (Alberta) s 108(1) and (2).

[86] For example, in Ontario, it is an offence to fail to comply with a term of a certificate of approval or to breach an administrative order. Environmental Protection Act (Ontario) s 186(3) and s 186(2) respectively.

[87] Office for National Statistics, *UK 2005* (London) 3.

[88] *Ibid.*

[89] Information from <http://www.statistics.gov.uk/CCI/nugget.asp?ID=6> accessed 5 January 2009. Over 90% of the UK population lives in urban areas, despite the fact that urban areas make up only 9% of the total land area.

> [T]he dematerialisation of the economy, along with changes in energy supply and an increasingly mobile lifestyle, has helped reshape UK production and consumption patterns in ways that eased some traditional pollution pressures and raised some new environmental challenges.[90]

New challenges include tackling diffuse pollution, waste production, climate protection and landscape conservation. In many of these areas, the UK has a reasonably encouraging recent environmental record. For example emissions of the basket of greenhouse gases covered by the Kyoto Protocol were 20.6 per cent lower in 2006 than in the baseline year of 1990.[91] Although much of this can be put down to the decline of industry and the associated reduction in the use of coal-fired power stations, the UK government is on course to meet and indeed surpass its Kyoto target of a 12.5 per cent reduction: in 2010, emissions are predicted to be 23.6 per cent below base levels.[92] There has also been a dramatic reduction in the occurrence of serious pollution incidents with 910 incidents occurring in 2006, the lowest number on record.[93] River quality has also improved with 71 per cent of rivers in England in 2006 being of good biological quality, a rise of 11 per cent from 1990 levels.[94] One final example of improvement relates to the state of land designed as a Site of Special Scientific Interest. In 2006, 72 per cent of these sites were in a favourable or recovering condition, a rise from 2003 levels.[95] In David Boyd's study of country environmental performance using OECD indicators, the UK was 13th overall, 14 and 15 places higher than Australia and Canada respectively.

The UK's environmental record is however far from unblemished. For example, although mineral, construction and demolition wastes, which account for over 70 per cent of total waste arisings in the UK have remained relatively constant in recent years, the amount of waste generated by households and business between 1995/96 and 2005/06 has increased by approximately 14 per cent. Saying that recycling and composting of household waste materials has risen from 6 per cent in 1995/6 to just over 25 per cent in 2005/06.[96] Other problems include historic contamination, air quality and electricity consumption.

B The Institutional Framework

What follows is a brief overview of the institutional framework for environmental protection in England and Wales. This framework has been complicated some-

[90] OECD, *Environmental Performance Reviews: United Kingdom—Conclusions and Recommendations* (2002) available at: <http://www.oecd.org/dataoecd/39/17/2395896.pdf> accessed 5 January 2009.

[91] DEFRA, *UK Climate Change Programme—Annual Report to Parliament July 2008*, 19.

[92] <http://www.defra.gov.uk/environment/climatechange/uk/progress/index.htm> accessed 5 January 2009.

[93] DEFRA, *Environment in Your Pocket 2007* (November 2007) (available at: <http://www.defra.gov.uk/environment/statistics/eiyp/index.htm> accessed 5 January 2009) 22.

[94] *Ibid* 38.

[95] *Ibid* 59.

[96] *Ibid* 51.

what by Welsh devolution. The Government for Wales Act 1998 established a National Assembly for Wales which debates and approves legislation and holds the Welsh Assembly Government to account. Although the Welsh Assembly has the power to pass secondary or delegated legislation, Westminster has retained the power to pass primary Acts of Parliament. As a result, the legal framework of environmental controls remains the same, but the Welsh Assembly can (and indeed does) use its law-making powers to tailor legislative controls to meet distinctive Welsh needs. Whilst devolution in Wales (and indeed in Scotland and Northern Ireland) has had some impact on pollution control regulation and the administration of environmental protection in the UK, there is a distinct lack of diversity across the UK. This can, in the main, be put down to the fact that much of the UK's environmental law and policy is shaped by European and international commitments and it is the UK which is both a member of the EC and a recognised nation state at international level. This undoubtedly constrains the extent to which environmental laws vary.

i Central Government

Since 1970, there has been a Department of the Environment although more recently, responsibility for the environment has been shared with a number of other policy areas such as local government and transport. The main government department currently responsible for environmental protection in England and Wales (in fact, the UK) is the Department of Environment, Food and Rural Affairs (DEFRA), created in 2001.[97] As is implicit in its name, environmental protection (including wildlife and countryside conservation) is merely one of DEFRA's remits. Other responsibilities include food and drink, animal health and welfare and horticulture. Its overarching mission is to 'enable everyone to live within environmental limits' and in doing so, it emphasises the need to tackle climate change through action at both international and domestic level and to 'secure a healthy, resilient, productive and diverse natural environment.'[98] As is the case in Australia and Canada, there are a number of substantial environmental policy areas which fall outside the remit of DEFRA. One of the most important is town and country planning, which is now in the hands of the Department of the Communities and Local Government. Other important government 'players' in environmental protection issues include the Department for Business, Enterprise and Regulatory Reform[99] (eg, producer responsibility and energy production) and the Department of Transport (eg, road building and airport expansion). One of the clear and inevitable difficulties of fragmenting responsibility in this way is that of finding integrated solutions to environmental problems. In recognising this, the government established, in 1990, the

[97] For a comprehensive overview of the administration of environmental law and policy at central government level see further S Bell and D McGillivray, *Environmental Law*, 6th edn (Oxford, OUP, 2006) ch 5.

[98] <http://www.defra.gov.uk/corporate/index.htm> accessed 5 January 2009.

[99] Formerly the Department of Trade and Industry (DTI).

Greening Government Initiative which required inter alia government departments to consider the environmental impacts of their policies and programmes. Rebranded in 2001 as 'Sustainable Development in Government', the current policy reflects sustainable development thinking and requires that social and economic impacts also be taken into account in government decision-making generally.[100]

The main role of DEFRA, along with all other government departments, is to advise ministers and implement government policies. Ministers or Secretaries of State also have very wide legislative and quasi-legislative powers. For example, a primary Act of Parliament, which normally provides a general legislative framework, may give Ministers the power to make 'delegated' or 'secondary' legislation that contains detailed legislative provisions on matters such as waste management and pollution control. There is no doubt that the work of DEFRA has an enormous impact on achieving environmental policy objectives. But Bell and McGillivray make three important qualifications to this statement.[101] First, they question the extent to which DEFRA is a particularly strong government department, especially in light of the fact that the previously constituted Department of Environment, Transport and Regional Affairs (DETR) had a much more significant portfolio of environmental responsibilities including town and country planning.[102] Planning and development control policy has a major impact on environmental protection and nature conservation and in my view should be a DEFRA responsibility. Secondly, they point to the fact that DEFRA has very few operational powers and those that it does have are often delegated to other bodies. Finally, they argue that DEFRA has an extremely wide portfolio of responsibilities 'some of [which] . . . not only take priority over environmental protection but are often in direct conflict with environmental aims.'[103]

In Wales, the Department for Environment, Sustainability and Housing is responsible for inter alia matters of environmental protection relevant only to Wales. Overall policy direction is provided by the Department's 2004 Wales Environment Strategy, with specific guidance on issues such as waste, water quality and climate change.

ii Regulatory Agencies—The Environment Agency

Prior to 1995, the responsibility for environmental regulation and policy was divided along sectoral lines with Waste Regulation Authorities (part of local authorities), the National Rivers Authority (NRA) and Her Majesty's Inspectorate of Pollution (HMIP) being responsible for discharges to land, water and air. By the

[100] For a critical overview see House of Commons Environmental Audit Committee Sixth Report, *The Greening Government Initiative 1999* (Session 1998–99) and A Ross-Robertson, 'Greening Government—Tales from the New Sustainability Watchdog' (2000) 12 *JEL* 175.

[101] Above n 97, 113–14.

[102] 'Restructure in haste, repent at leisure' (2001) 317 *ENDS Report* 2.

[103] Above n 97, 114. For example, the regulation of the economic and competition aspects of the water industry.

mid to late 1980s, calls for an integrated, cross-sectoral environmental protection agency were strong.[104] Although initially rejected by the government of the time, in 1991 there was an important reversal of policy and the new Secretary of State, Michael Heseltine agreed to 'look at these matters afresh.'[105] The details of this new 'integrated' Agency emerged slowly. According to Bell and McGillivray, there were two key sticking points.[106] The first related to whether the entire remit of the NRA, which was extensive and included not only pollution control, but matters such as water resource management, fisheries and navigation, should be transferred to an integrated Agency. The second concerned the waste regulation duties of local authorities and whether they should be included within the agency's ambit: uniformity would be increased but at the expense of local accountability. Finally, after much consultation, the government announced that the new Agency would take over all the functions and duties of the NRA and HMIP, along with the waste regulation duties of local authorities.

The Agency, a Non-Departmental Public Body of DEFRA and an Assembly Sponsored Public Body of the National Assembly for Wales, was established in the Environment Act 1995 and became operational on 1 April 1996. It is the second largest agency of its type in the world, and the largest in Europe.[107] It employs over 13,000 staff to carry out its duties and functions and in 2006–07 had an operational budget of approximately £1.002 billion, about 60 per cent of which (£603 million) was grant from the Agency's sponsoring departments. The rest of the Agency's funding is raised from various charging schemes and flood defence levies. These schemes are subject to the principle of cost recovery meaning that charges must be fixed to recover the costs incurred on the activity to which the charges relate. Despite the Agency's relatively large operational budget, less than 30 per cent is spent on environmental protection activities such as waste regulation, process industries regulation and land quality. Over 50 per cent of the budget is ring-fenced for flood defence activities.[108]

Although the Agency is by no means the only authority that has control over environmental regulation, with a number of responsibilities such as drinking water, nature conservation and landscape protection somewhat surprisingly falling outwith its remit, it is undoubtedly the most important in terms of pollution control regulation.[109] Its functions and responsibilities are innumerable. It regulates a large number of processes and activities that have an impact on the

[104] For an overview of the background to the establishment of the Environment Agency see N Carter and P Lowe, 'The Establishment of a Cross-Sector Environment Agency' in TS Gray (ed), *UK Environmental Policy in the 1990s* (London, Mcmillan Press, 1995) 4.

[105] House of Commons Environment Committee (1991–92) *The Government's Proposals for an Environment Agency*, Session 1991–92 First Report (London: HMSO) 1.

[106] Above n 97, 124.

[107] Select Committee on Environment, Food and Rural Affairs Seventh Report 2005–06, *The Environment Agency* HC (11 May 2006) 780–I [1].

[108] *Ibid*, Memorandum submitted by the Environment Agency.

[109] Other important environmental 'players' include the Drinking Water Inspectorate and Natural England and the Countryside Council for Wales which are responsible for inter alia protecting designated areas and promoting access to the countryside.

environment, including the abstraction and storage of water, salmon and freshwater fisheries. It is also largely involved with the construction and maintenance of flood defences, is responsible for the running of a flood-warning service and is a statutory consultee in relation to planning applications. All in all, the Agency's remit covers 15 million hectares of land, 36,000 kilometres of river and 5,000 kilometres of coastline.[110] The overriding aim of the Agency in discharging all its functions is to make the contribution towards attaining the objective of achieving sustainable development.[111] Subject to this aim, the Agency has an array of secondary duties. So for example, the Agency's pollution control powers are exercisable for the purpose of preventing, minimizing, remedying or mitigating the effects of pollution[112] and, with respect to water, the Agency should promote the conservation and enhancement of the natural beauty and amenity of inland and coastal waters and of land associated with such waters.[113]

As I have already noted, the Agency is classed as a Non-Departmental Public Body (NDPB) or 'quango', ie, it has a role in the processes of government but is not a government department or part of one and accordingly, to some extent, operates at arm's length from ministers. The arguments for the establishment of an 'independent' body seem very persuasive. O'Riordan and Weale, for example, declare that 'independence of regulation both from those who are licensed and from governments that might interfere' is a fundamental principle on which the greening of the machinery must be found.[114] The need for independence was echoed in the Select Committee on the Environment's Report on Toxic Waste where it was felt that environmental protection should not be subject to the vagaries of political fortune and the need to maintain public and scientific confidence demanded objectivity.[115] Unlike government departments, independent agencies can also provide a quick and flexible response to matters of particular concern as they are not constrained by the formal parliamentary process. This is particularly apt in the environmental sphere, for environmental hazards may require immediate attention, environmental damage may require immediate remediation and both may necessitate a speedy change in management focus and policy. Furthermore, it could be argued that, by being distanced from parliament and the executive, agencies are more able to encourage the participation of the public and interest groups in its decision-making processes. But delegation of power to an Agency does not necessarily entail that the agency is free from political interference. In fact, under the Environment Act 1995, DEFRA and its ministers retain considerable control over Agency activities by, inter alia, directing

[110] <http://www.defra.gov.uk/ENVIRONMENT/ea/index.htm> accessed 5 January 2009.
[111] Environment Act 1995 s 4(1).
[112] Environment Act 1995 s 5(1).
[113] Environment Act 1995 s 6(1)(a).
[114] T O'Riordan and A Weale, *Greening The Machinery of Government* (London, Friends of the Earth, 1992) para 2.11, cited in Carter and Lowe above n 104, 44.
[115] Cited in Carter and Lowe *Ibid* 44. Second Report of the Select Committee on the Environment, *Toxic Waste* HC (1988–89) 22.

Environment Agency action, appointing Board members, determining levels of government sponsorship and drafting guidance.

The transition from a fragmented administrative approach to environmental protection to the creation of an integrated 'environmental' regulator in 1996 has not been flawless. Ten years on, the House of Commons Environment, Food and Rural Affairs Select Committee investigated the workings and operation of the Agency and in doing so, identified a number of key issues.[116] For example, one of the most fundamental concerns related to the Agency's wide range of responsibilities which require it to act as a regulator, a 'champion of the environment' and an adviser to business. The Committee identified tensions between these various functions and recommended the need for greater clarity.[117] Inconsistency of approach (either between geographical regions or between the policy centre and inspectors on the ground) was also identified by stakeholders as a serious concern. Although real progress has been made since the problem was first identified in 1999, the Committee reiterated the need for greater communication between the policy centre and the inspectors on the ground.[118]

iii Local Government

Local authorities in England and Wales also play an important role in furthering environmental protection objectives. One of the most important functions of local authorities is to manage the system of town and country planning. In most cases, they are the local planning authority for their particular area and are therefore responsible for making development plans and controlling development. Other planning-related matters include tree preservation orders, conservation areas, listed building protection and hazardous substances consents. In addition to planning control, they are responsible for inter alia identifying and securing the remediation of contaminated land, collecting and disposing waste, controlling noise from premises and regulating air pollution emanating from certain types of industrial premises.[119] Sustainable development is at the core of their regulatory functions, with section 4 of the Local Government Act 2000 placing on all local authorities a duty to prepare a community strategy for:

> [P]romoting or improving the economic, social and environmental well-being of their area and contributing to the achievement of sustainable development in the United Kingdom.

[116] Above n 107.
[117] *Ibid* paras 13–25.
[118] *Ibid* paras 32–8.
[119] For a brief overview see Bell and McGillivray above n 97, 132–3.

C The Legislative Framework

The UK arguably has the oldest system of pollution control laws in the industrialised world. From a regulatory perspective, it can be traced back to medieval statutes tackling small-scale pollution. But until the mid- to late-19th century, it was the development of private law (common law) principles which afforded some protection to communal assets such as water. These principles were not however concerned with protecting the 'environment' *per se* and furthermore, their ability to protect even private, individual interests was somewhat limited. As identified by the Royal Commission on the Pollution of Rivers in 1867, 'the expense of such litigation generally far exceeds the value of the personal interest of any individual'.[120] The beginnings of public law controls specifically related to environmental protection originate in the Industrial Revolution, when the emergence of health problems such as typhoid and cholera, both of which were clearly linked to an unclean urban environment and the drinking of contaminated water, led to the passing of public health controls. The Public Health Act of 1872 imposed, for the first time, mandatory duties on local sanitary authorities and the legislation was consolidated in the Public Health Act 1875. According to Bell and McGillivray, although these Acts were largely successful in removing sewage from industrial town centres, this success was achieved by facilitating the disposal of untreated waste into local watercourses.[121] In response, the Rivers Pollution Prevention Act 1876, the first statute to deal with water pollution was passed and:

> [G]ives a flavour of the traditional approach to pollution control through the use of criminal offences and a practical reliance on the 'best practicable means' of pollution control.'[122]

Around the same time, in 1863 the government created the Alkali Inspectorate to control atmospheric emissions produced by the caustic soda industry. Again in response to public health problems linked with the industrial revolution, the controls relating to town planning were introduced in the Housing, Town Planning etc Act 1909 although it was some 40 years later, still early in world terms, that obligatory planning controls were introduced in the Town and Country Planning Act 1947. As a result of this evolutionary process, environmental laws could, until recently, be characterised by several core features. One such feature was that 'government structures and legislation have been a fragmented accretion of common law, statutes, agencies, procedures and policies', enacted in response to individual problems.[123] Other key characteristics included the extensive use of uniform rather than flexible emissions standards and an emphasis on negotiation and persuasion in enforcing environmental controls.

[120] Royal Commission on the Pollution of Rivers, Third Report, *The Rivers Aire and Calder* (1867) Cmnd 3850 pp ii–iii, cited in Bell and McGillivray *Ibid* 20–21.

[121] *Ibid* (Bell and McGillivray) 21.

[122] *Ibid.*

[123] Above n 104, 40.

It is true to say that some of these features are still in evidence today. But in the 1970s, when there was increasing pressure on the state to enhance its role in controlling pollution and protecting the environment, the UK witnessed the growth of a 'modern age' of environmental law.[124] The process of change began in 1970 itself with the creation of the Department of the Environment charged solely with the responsibility of advancing environmental protection objectives. There followed both a rapid growth in the range of environmental controls and a fundamental change in regulatory approach. This can be traced back to the Control of Pollution Act (COPA) 1974 which, whilst not going so far as to introduce an integrated approach to pollution control, targeted pollution caused by a number of industrial processes across a range of environmental media, and put most of the law on waste disposal and water pollution in one place. Although this Act has been virtually replaced by a number of important statutes, it paved the way for a more integrated and coherent approach to environmental regulation. A system of integrated pollution control was first introduced in 1990 in Part I of the Environmental Protection Act (EPA), although this has now been replaced by the Pollution Prevention and Control Act 1999. The establishment of the Environment Agency in 1995 continued this integrated approach to environmental problems. Other main environmental statutes include Part II of the EPA, which regulates the production, treatment and disposal of controlled waste and the Water Resources Act 1991, which controls discharges into controlled waters.

Current pollution control statutes do, in the main, follow a 'command and control' model whereby environmental standards are set centrally and enforced by a combination of government and regulatory agencies. They are underpinned by criminal liability, although other administrative-based enforcement mechanisms can be used to deter non-compliance. Despite a more recent emphasis on combining command and control approaches with alternative regulatory techniques such as market-based mechanisms and voluntary instruments, the command and control model still forms, as it does in Australia and Canada, the backbone of pollution-control law in England and Wales.

One also cannot overemphasise the important role of the European Union in shaping modern-day environmental legislation.[125] Since the early 1960s, the EU has proved a key player in legislating across a wide range of environmental issues, from industrial pollution control, to protection of biodiversity and waste minimisation. Like most jurisdictions, the EU is experimenting with alternative regulatory instruments such as taxes, subsidies and trading scheme, and voluntary measures.[126]

[124] For an excellent summary of developments in UK environmental regulation, see Bell and McGillivray above n 97 ch 2.

[125] For an introduction to EU environmental law see M Lee, *EU Environmental Law: Challenges, Change and Decision-Making* (Oxford, Hart Publishing, 2005) esp ch 1, and J Holder and M Lee, *Environmental Protection Law and Policy: Text and Materials*, 2nd edn (Cambridge, CUP, 2007) ch 4.

[126] See eg Directive 2003/87/EC of 13 October 2003 establishing a scheme for greenhouse gas emission allowance trading within the Community [2003] OJ L275/32 and Regulation 761/2001 of 19 March 2001 allowing voluntary participation by organisations in a Community eco-management and audit scheme (EMAS) [2001] OJ L114/1.

However, licensing or permitting systems remain at the heart of EC environmental legislation.[127] The majority of EU environmental law is in the form of Directives, which have to be implemented by Member State governments. Ultimately, the European courts hold national governments to account for both practical and formal implementation.

V Conclusion

In this chapter, I have summarised some of the environmental pressures faced by governments in Australia, Canada, and England and Wales and introduced the institutional and regulatory frameworks designed to advance environmental protection objectives. In all three countries, environmental protection and pollution control issues remain high on the political agenda. Although governments can demonstrate some success in reducing pollution and tackling more 'modern' environmental concerns such as climate change and loss of biodiversity, there are still significant challenges ahead.

The chapter has emphasised the fact that environmental responsibilities in Australia, Canada, and England and Wales are spread across several layers of government. In Australia and Canada, federal government sets the policy framework and has a key role in legislating on national environmental issues. Implementing and enforcing these environmental provisions rests with a division within the relevant government department. Provincial, state and territorial governments are equally if not more important in furthering environmental protection objectives within their geographical boundaries. They are responsible inter alia for regulating polluting processes and controlling emissions to air, water and land. In some states such as Victoria (Australia), the responsibility for monitoring and enforcing compliance with environmental law is vested in a statutory authority that operates at arm's length from the government department. Local governments also have a range of environment-related responsibilities. In order to ensure inter-jurisdictional consistency, both Australia and Canada have well-established bodies charged with the responsibility of ensuring adequate and effective co-operation between different layers of government. In England and Wales environmental responsibilities are divided between a central government department (DEFRA) and its executive agency (Environment Agency), and local authorities. In all three countries, there are a variety of political, social, economic and practical difficulties faced by environmental departments and regulators.

Despite the institutional variations, in all three countries, 'command and control' regulation is at the core of the legislative framework. Prior approval (secured by the issuing of a licence or permit) is required for certain regulated activities and

[127] See, eg, Directive 2008/1/EC of 15 January 2008 concerning integrated pollution and control (a codified version) [2008] OJ L24/8.

operations are controlled through the use of licence conditions incorporating inter alia environmental standards and operating requirements. Non-compliance with the regulatory controls is underpinned by criminal liability and regulators can utilise a variety of compliance and enforcement mechanisms in responding to violations.

Now that we have an understanding of the institutional and legislative frameworks relevant to pollution control law and its enforcement in Australia, Canada, and England and Wales, we can move on to consider, in Chapter 5, the enforcement strategies adopted by environmental regulators when enforcing pollution control regulation. Chapters 6, 7 and 8 then examine in detail, the enforcement mechanisms available to regulators, their deterrence characteristics and the extent to which they promote the goal of cost-effective enforcement, as discussed in Chapters 2 and 3.

5

Environmental Enforcement Strategies in Australia, Canada, and England and Wales

I Introduction

AS DISCUSSED IN Chapter 3, the enforcement strategies of regulatory authorities can vary considerably. The traditional distinction made in the academic literature between compliance and deterrence-based approaches has in many respects been abandoned and the scholarly debate has moved forward to acknowledge the effectiveness of more sophisticated regulatory techniques which may reflect both compliance and deterrence characteristics. Responsive regulation is but one example of such an approach. This chapter examines the enforcement strategies of pollution control regulators in parts of Australia and Canada, and England and Wales. Although there is considerable variation in the regulatory approach to enforcement, these regulators generally reject a pure deterrence approach to enforcement. Instead, they adopt a more flexible and conciliatory strategy which emphasises informal enforcement action (such as education and persuasion) but at the same time acknowledges the important role that prosecution and other formal sanctions can play in promoting a compliance agenda.

But do current enforcement practices send a strong enough deterrence signal? Are these more flexible enforcement strategies underpinned by a real and credible deterrence threat? As noted above, formal enforcement action tends to be the exception rather than the rule. Prosecutions are relatively rare and, although other formal enforcement tools such as administrative notices and monetary administrative penalties are a more likely enforcement response, the majority of incidents of non-compliance are dealt with by informal enforcement techniques such as persuasion, education and informal warnings. So how does this impact on the deterrence value of enforcement? The deterrence framework advanced in Chapter 2 predicts that enforcement will induce compliance where the benefits derived from non-compliance are less than the expected sanction, as determined by the probability and associated costs of an enforcement response. Where, as in our case, the probability of formal enforcement action is low, under the expanded deterrence framework (as summarised in Chapter 2 Figure II) the values of 'e' (the probability of administrative enforcement action) and 'c' (the probability of formal court action) are small, thus discounting the deterrent impact of any formal enforcement sanction. Of course, formal enforcement action, particularly criminal prosecution is expensive. The key then, is to promote a high level of compliance by maximising the deterrent potential of enforcement action that is taken.

I consider the enforcement strategies of environmental regulators in Australia, Canada, and England and Wales[1] with a view to determining the place of formal, deterrence-based sanctions in the enforcement process. I use available enforcement statistics to evaluate the extent to which my chosen regulatory authorities use deterrence-based mechanisms such as, enforcement notices and towards the top of the 'enforcement pyramid', coined as such by Ayres and Braithwaite, criminal prosecution and licence/permit suspension and revocation.[2] The lower an enforcement mechanism is in the pyramid, the more indicative it is of a persuasive approach to enforcement. I also make reference to published enforcement policies which not only provide a valuable insight into enforcement approaches, but also go some way to constraining the broad discretionary powers afforded to regulatory authorities.[3]

II Australia

A Introduction

Historically, the enforcement approach of environmental regulators in Australia has been one of compliance as opposed to deterrence. Grabosky and Braithwaite, in their now dated study of the enforcement strategies of Australian business regulatory agencies in 1984/85, observed that 'despite the wide variations across Australia in policies relating to prosecution, environmental regulators invariably seek co-operative relationships with industry.'[4] This trend has continued and Australian environmental regulators generally adopt a predominantly compliance-based approach to enforcement, treating criminal prosecution as a last resort. Take the states of Queensland and South Australia as examples where, as indicated in Table I below, prosecution numbers are low.

[1] For an overview of selected environmental regulators in Australia (federal, NSW and Victoria), Canada (federal and Ontario) and England and Wales, see generally ch 4.

[2] I Ayres and J Braithwaite, *Responsive Regulation* (Oxford, OUP, 1992). See further ch 3 s III B for a discussion of responsive regulation and the enforcement pyramid.

[3] For a brief insight into the importance of discretion in the enforcement process see ch 3 s III B.

[4] P Grabosky and J Braithwaite, *Of Manners Gentle: Enforcement Strategies of Australian Business Regulatory Agencies* (Oxford, OUP, 1986) 47.

Table I—Prosecution Statistics: Queensland and South Australia[5]

	2005–06	2004–05	2003–04	2002–03	2001–02
EPA South Australia *	6	3	6	2	5
EPA Queensland **	7	2	7	7	7

* finalised prosecutions
** prosecutions under the Environmental Protection Act 1994

In contrast, pollution control authorities in NSW and Victoria have utilised a more prosecutorial approach to enforcement than environmental regulators in other Australian states and territories.[6] However, despite being described as agencies that have 'regulatory strategies of moderately strict enforcement'[7] and regulators that place a 'greater emphasis . . . on prosecution',[8] it would be misleading to describe these regulators as adversarial in their enforcement activities. This section begins by examining the enforcement approach of the Department of the Environment, Water, Heritage and the Arts (DEWHA) at federal level. It then considers the pollution control enforcement activities of the Victorian Environmental Protection Authority (EPA (Victoria)) and the Department of Environment and Climate Change (DECC) in NSW.

B The Department of Environment, Water, Heritage and the Arts (DEWHA)

Although the approach of the Commonwealth DEWHA (in conjunction with the Australian Federal Police and Australian Customs Service) in enforcing the provisions of the Environment Protection Biodiversity and Conservation Act (EPBCA) 1999 varies depending on the nature of the breach, broadly speaking, a compliance approach is preferred:

> Resorting to legal action to enforce the Act is not undertaken lightly and used only in the most serious cases, and only after co-operative or alternative approaches to resolving disputes have failed.[9]

Indeed the most recent Compliance and Enforcement Policy produced by the Department of Environment and Heritage (the predecessor to the DEWHA)

[5] Statistics taken from Annual Reports.

[6] See eg J Norberry, 'Australian Pollution Laws: Offences, Penalties and Regulatory Agencies' in Australian Institute of Criminology Conference Proceedings No 26, *Environmental Crime: Proceedings of a Conference held 1–3 September 1993*, 10.

[7] Above n 4, 38.

[8] B Robinson, *Review of the Enforcement and Prosecution Guidelines of the Department of Environmental Protection of Western Australia* (February 2003). See also Grabosky and Braithwaite (above n 4) and S Streets, 'Prosecuting Directors and Managers in Australia: A Brave New Response to An Old Problem?' (1998) 22 *Melbourne University Law Review* 693, 698.

[9] DEWR Press Release, 'Record penalty for illegal clearing of wetland' (14 October 2004).

emphasises the importance of compliance-based enforcement approaches such as communication and education, persuasion, co-operative assistance and collaboration.[10] But where these approaches fail, other enforcement mechanisms may need to be used and a responsive approach will be adopted:

> The Department employs a range of responsive enforcement sanctions that escalate in severity as the need arises. These sanctions rely on the deterrent effect of penalty-based instruments such as suspension or cancellation of permits, injunctions, remediation orders, pecuniary penalties, and criminal prosecution.[11]

Responses will escalate according to the severity of the contravention or if non-compliant activities continue.[12] Seriousness of harm, likelihood of repeat offence, history of the offender and cost to the Government or general community of the contravention are just four of the criteria against which the severity of an offence should be judged.[13] The policy clearly states that first and less serious offences should be dealt with through education and/or warnings. It is only for serious or continuing offences that deterrence sanctions such as remediation orders, pecuniary penalties and criminal prosecution should be used.[14]

The policy divides enforcement responses into three categories: administrative action, civil action and criminal action. Administrative actions should be used for relatively minor contraventions or where the suspect has been especially co-operative.[15] Somewhat surprisingly, the policy appears to consider licence suspension and revocation as appropriate for minor contraventions. These two sanctions are generally viewed as being at the top of the enforcement pyramid. In the case of serious civil contraventions[16], civil sanctions such as court orders for repair and mitigation of damage may be sought. Criminal action will only be taken for serious criminal offences and the Department must have sufficient evidence to prove both the physical and fault-based element of the offence beyond reasonable doubt. A serious criminal offence is one which displays certain characteristics such as a significant degree of criminality (or a repeat offence) and the risk of significant actual or potential harm to the environment.[17]

In terms of enforcement of the EPBCA at ground level, where, for example, the offence relates to non-compliance with the environmental approvals process, the

[10] DEWHA, *Compliance and Enforcement Policy* (May 2008) 4 available at <http://www.environment.gov.au/about/publications/pubs/compliance-enforcement-policy.pdf> accessed 5 January 2009.

[11] *Ibid* 5.

[12] *Ibid* 7.

[13] *Ibid* 8.

[14] *Ibid* 7.

[15] *Ibid* 9.

[16] The EPBCA creates civil (as well as criminal) penalty provisions. For example, taking, without approval, action that is likely to have a significant impact on matters of national environmental significant is a civil contravention (ss 12 and 142 EPBCA). According to the policy (*Ibid* ss 9–10) a 'serious' civil contravention has at least one of a list of attributes. For example, a serious civil contravention will involve 'a blatant disregard for or significant degree of indifference to the civil law' and 'the Australian Government or the community expects that the matter will be dealt with by way of enforcement action.'

[17] *Ibid* 10.

DEWHA will rarely prosecute, with only one prosecution completed in 2005–06 and two matters referred to the Director of Public Prosecutions in 2006–07.[18] In contrast, prosecutions for breach of the provisions relating to international wildlife trade are, comparatively speaking, much more frequent. For example in 2005–06, 15 charges were laid against 12 defendants for breaching the international wildlife trade provisions contained in Part 13A of the Act, although 5,165 seizure notices were issued for international wildlife crime. Similarly in 2006–07, two charges were laid and 7,533 seizure notices were issued.[19]

One should, however, be cautious in comparing federal enforcement with that of the states and territories for as noted earlier, the Australian Commonwealth government and its institutions have a minimal role to play in pollution prevention. Nonetheless, the enforcement strategy of the DEWHA is interesting, not least because of recent changes made to the EPBCA which reflect the difficulties the department has experienced in prosecuting companies and individuals for non-compliant behaviour.[20] In addition to introducing a new range of enforcement options which can act as an alternative to lengthy and expensive court proceedings, the amended Act strengthens the role of the criminal law by for example applying strict liability to a range of offences relating to inter alia matters of national environmental significance, Commonwealth land and the taking, injuring or killing of listed threatened species.[21] Such changes arguably reflect the DEWHA's recognition of the importance that prosecution (or the threat of it) plays in securing compliance.

C Department of Environment and Climate Change (DECC) (New South Wales) and Environmental Protection Authority (EPA) (Victoria)

The DECC and the EPA Victoria produce enforcement and prosecution guidelines which summarise the principles to be followed in relation to enforcement activity. In both cases, the guidelines clearly reflect a compliance-based enforcement strategy, but at the same time deem prosecution to be appropriate in some cases.

The Enforcement Policy of the EPA Victoria recognises that non-regulatory measures are 'often effective and reduce the need for enforcement', such measures to include education, the provision of information and technical advice on licence compliance and waste minimisation.[22] However, where an offence is committed,

[18] See Operation of Environmental Protection and Biodiversity Conservation Act 1999 Annual Reports 2005–06 and 2006–07 (available at: <http://www.environment.gov.au/about/publications/annual-report/06-07/pubs/vol2-environmental-protection.pdf> accessed 5 January 2009).

[19] *Ibid.*

[20] Environment and Heritage Legislation Amendment Act (No 1) 2006 (No 165).

[21] See eg Environment and Heritage Legislation Amendment Act (No 1) 2006 (No 165) ss 24, 26 and 28.

[22] EPA Victoria, *Enforcement Policy* (July 2006) 8–9 available at: <http://epanote2.epa.vic.gov.au/EPA/Publications.NSF/2f1c2625731746aa4a256ce90001cbb5/39a0493666adf794ca2571a2002cdb9e/$FILE/384.3.pdf> accessed 5 January 2009.

enforcement action will be taken, which 'may or may not involve prosecution.'[23] Deciding which enforcement measures are most appropriate in a given set of circumstances is a matter of judgement based on a range of factors including the seriousness of the offence, the cost of enforcement, the culpability of the offender and the extent to which the offender co-operated with the regulator.[24] According to the Policy, regulatory enforcement is a 'staged process' involving a hierarchy of enforcement options. This is represented in Figure VI below.

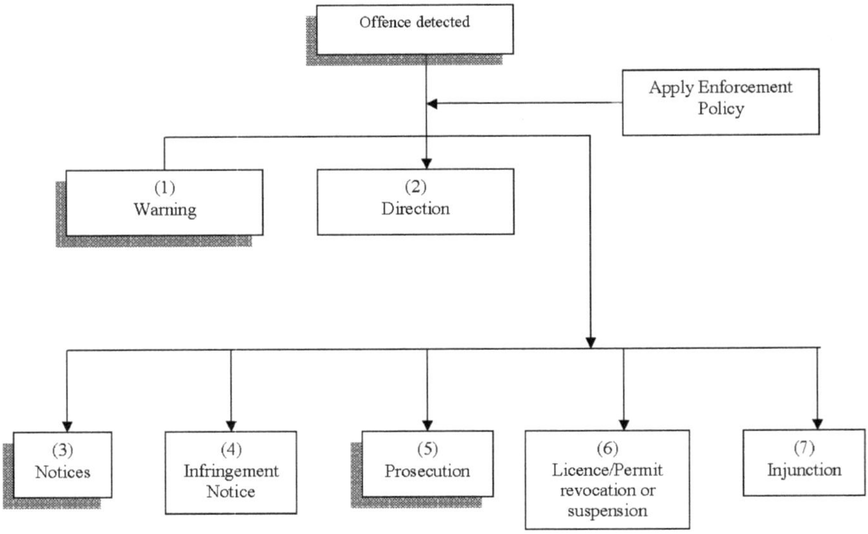

Figure VI: Environment Protection Act—Enforcement Decision-Making[25]

Prosecutions will only be brought where the available evidence establishes a prima facie case and where, for example, the case involves serious harm or risk to the environment, repeat offences or deliberate non-compliance. Prosecution is also recommended where an offender has failed to comply with previously instituted enforcement action such as non-compliance with a direction or a penalty infringement notice.[26] Licence suspension/revocation is clearly (with injunctions) at the apex of the pyramid thereby reflecting the seriousness of its consequences. This is in line with Ayres and Braithwaite's enforcement pyramid.[27]

[23] *Ibid* 10.

[24] *Ibid* 10 and app 1.

[25] *Figure VI* is based, closely on a diagram contained in the EPA (Victoria) Enforcement Policy. *Ibid* 15.

[26] *Ibid* 12.

[27] See ch 3 s III B.

The Prosecution Guidelines of the DECC in New South Wales (drafted by the regulator at the time, the Environment Protection Authority) and a review of the Protection of the Environment Operations Act (PEOA) 1997 carried out in 2005, also refer to a compliance approach, whilst, at the same time, highlighting the importance of prosecution.[28] The guidelines indicate that not every breach of the criminal law will lead to automatic prosecution, describing the laying of charges as 'discretionary'.[29] According to the Policy, the EPA (DECC):

> [M]ay pursue non-prosecution options to prevent, control, abate or mitigate any harm to the environment caused by an alleged offence or to prevent the continuance or recurrence of an alleged offence. . . . Prosecution will be used, therefore, as part of the EPA's overall strategy for achieving its objectives. Each case will be assessed to determine whether prosecution is the appropriate strategic response.[30]

Prosecutions are therefore seen as one response to incidents of non-compliance although the guidelines and the PEOA review emphasise that they (and on-the-spot fines) 'remain a very important part of EPA's regulatory approach.'[31] A prosecution will, however, only be brought where it is in the public interest to do so, taking into account factors such as the history of the offender, the seriousness of the environmental impact, the culpability of the offender and the availability and efficacy of any alternatives to prosecution.[32] The guidelines themselves clearly (and perhaps unsurprisingly given their title) centre on the use of prosecution with very little mention being made of alternative enforcement mechanisms such as warnings and administrative notices. Although prosecution is discussed in the context of other enforcement mechanisms, the guidelines seem to place prosecution at the centre of the DECC enforcement strategy. This is somewhat in contrast with other Australian states (including Victoria) where the enforcement strategy is much less (overtly) reliant on prosecution. For example, in a recent review of the Enforcement and Prosecution Guidelines applicable in Western Australia, it was noted that the tone and language of those guidelines appeared to strongly discourage prosecution except where all other options were exhausted.[33] The question now is whether the enforcement statistics reflect the enforcement approach advanced by the DECC (NSW) and the EPA in Victoria.

As NSW is the most populated and heavily industrialised state in Australia, it is perhaps not surprising that prosecution numbers are highest here and the DECC (and its predecessor the EPA) has consistently been more active in prosecuting

[28] Department of Environment and Conservation/EPA, *EPA Prosecution Guidelines* (July 2004) available at: <http://www.environment.nsw.gov.au/legislation/prosguide/index.htm> accessed 5 January 2009. Department of Environment and Conservation (NSW), *Report on the Review of the Protection of the Environment Operations Act 1997* (October 2003) available at <http://www.environment.nsw.gov.au/resources/legislation/reportpeoereview.pdf> accessed 5 January 2009.

[29] *Ibid* (Prosecution Guidelines) para 3.2.

[30] *Ibid* paras 3.5–3.6.

[31] Above n 28 (Report on the Review), 11.

[32] Above n 28 (Prosecution Guidelines) para 3.7.

[33] Above n 8 (Robinson), 22. The report proposed a revision of the guidelines to make it clear that the prosecution policy is based on an appropriate response not the first or last response.

environmental offences than any other Australian jurisdiction.[34] The more aggressive enforcement approach began in the late 1980s when calls for tougher penalties against polluters were mirrored by legislative amendments designed to reflect the change in mood.[35] The Environmental Offences and Penalties Act 1989 allowed for higher penalties and in 1991, the State Pollution Control Commission, with its emphasis on self-control and self monitoring, was replaced by a revamped and more powerful Environment Protection Authority whose approach gave greater recognition to the role of prosecution. The cumulative effect of these legislative and administrative changes was a sharp increase in the number of environmental prosecutions with numbers rising from just 15 in 1984–85 to 54 in 1991–92.[36]

Today, the DECC in NSW clearly has a far more aggressive approach to prosecution than any other Australian state (including Victoria).[37] As illustrated in Figure VII below, the number of prosecutions completed under Environmental Protection Authority legislation has, however, declined since 2004–05. This reflects the lower number of court-elected infringement notices,[38] and the DECC's decision to target prosecutorial resources at serious breaches of legislation for threatened species, endangered populations and Aboriginal heritage provisions under the national Parks and Wildlife Act 1974. The DECC's continuing commitment to punishing environmental crime through prosecution is however evidenced by recent amendments to the PEOA 1997 (NSW) which inter alia strengthen provisions relating to offences and penalties. Under the Protection of the Environment Operations Amendment Act 2005, maximum penalties have been increased and alternative sentencing orders have been expanded.[39] Statistics would indicate that the availability of higher maxima is filtering through to the courts. Although the number of EPA prosecutions dropped, total financial penalties for EPA prosecutions rose from $642,650 in 2005–06 to $751,250 in 2006–07.[40]

In contrast, Figure VIII below would suggest that compared with the DECC (NSW), only a small number of major prosecutions are instigated by the EPA Victoria. These statistics are not, however, easy to interpret, mainly because EPA Annual Reports only stipulate the number of 'major' prosecutions (most of which

[34] T Howard, 'Liability of Directors for Environmental Crime: the Anything-but-Level Playing Field in Australia' (2000) 17 *Environmental and Planning Law Journal* 250, 263.

[35] For an excellent summary of state-wide enforcement activity in the 1990s see M Comino and P Leadbeter, 'Enforcement of Pollution Laws in Australia—Past Experience and Current Trends', paper presented at the Fifth International Conference on Environmental Compliance and Enforcement (November 1998) available at <http://www.inece.org/5thvol1/comino.pdf> accessed 5 January 2009.

[36] Statistics provided by Norberry above n 6, 7.

[37] See also Table 1 for prosecution statistics of two other Australian states: Queensland and South Australia.

[38] For detail see ch 8 s II C.

[39] For a summary, see <http://www.environment.nsw.gov.au/legislation/poeoreview.htm/poeoreview.htm> accessed 5 January 2009. See also ch 7 s III B i.

[40] Department of Environment and Climate Change, *Annual Report 2006–07*, 37

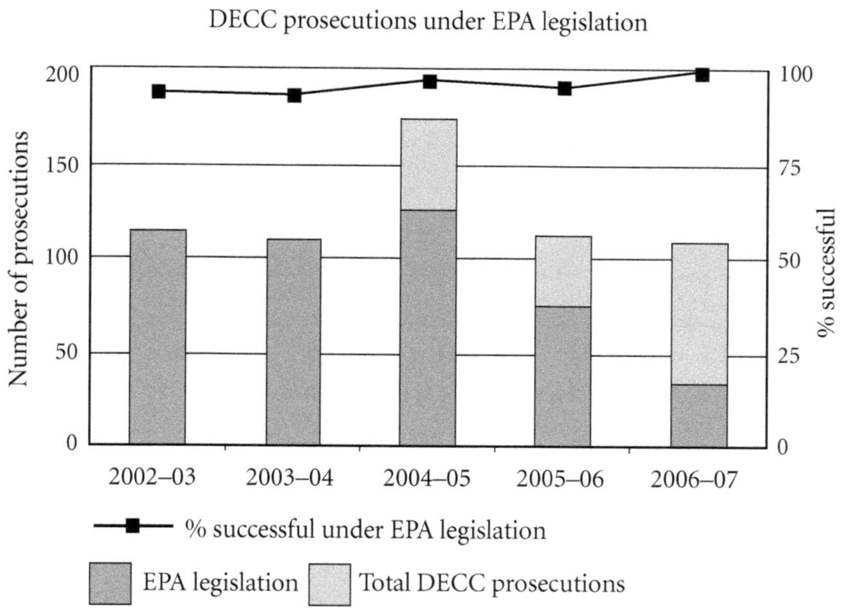

Figure VII: New South Wales—Department of Environment and Climate Change[41]

are brought under the provisions of EPA 1970) as opposed to the total number of prosecutions. Actual prosecution numbers may therefore be higher. Presuming that the EPA does indeed prosecute more frequently than is indicated by the data in its Annual Reports, it would seem that the enforcement approach of environmental regulators in NSW and Victoria (especially NSW) is more adversarial than in the other Australian states and territories.

However, it would be misleading to describe the enforcement policies and practices of the DECC and the EPA Victoria as reliant on prosecution. Compared with the use of other formal enforcement mechanisms such as abatement and infringement notices, the initiation of criminal prosecution is relatively rare. In 2006–07, for example, the EPA Victoria issued 197 pollution abatement notices and 44 clean-up notices, compared with 13 'major' prosecutions.[42] In the same time period, the DECC issued approximately 2,000 Penalty Infringement Notices for minor breaches of EPA legislation administered by the DECC compared with only approximately 33 prosecutions although of course, one would not normally expect minor breaches to attract prosecution.[43] Unfortunately, there appears to be no data on the use of other administrative enforcement mechanisms in NSW, thus making further comparisons difficult.

[41] Taken from Annual Report, *Ibid.*
[42] EPA (Victoria), *Annual Report 2006–07* app, 9, 12–22.
[43] Above n 40, 36.

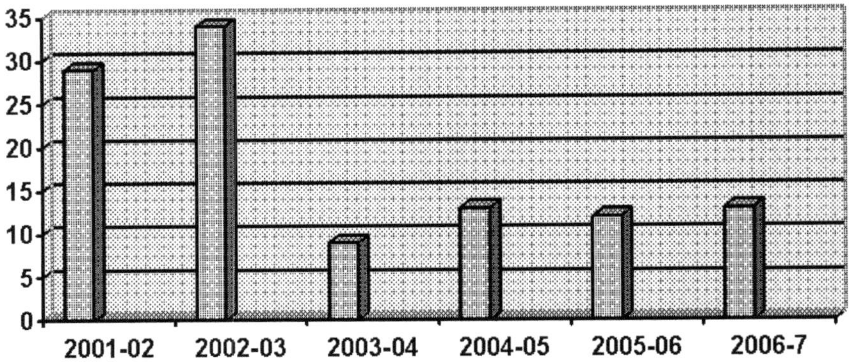

Figure VIII: Major Prosecutions Completed by the EPA Victoria[44]

It is clear that the DECC and EPA (Victoria) have rejected an enforcement strategy that places an emphasis on prosecution.[45] Although I have not been able to find statistics on the number of incidents of non-compliance, it would appear that in the majority of cases, an informal enforcement response will be used. During interviews with EPA (Victoria) officials, two specific reasons were given for this. First, maintaining a good relationship with the regulated community was regarded as important and as illustrated in Chapter 3, an overly penal approach to enforcement may have an adverse effect on regulatory co-operation.[46] Second, one official viewed the majority of Victorian businesses as good corporate citizens who try to do the right thing, with the majority of incidents resulting not from economic calculation but from ignorance, stupidity or human error. A deterrence-centred strategy may not be suitable when the regulated community displays such characteristics.[47] During an interview with the head of Legal Services in the EPA (now the DECC), the importance of a compliance strategy was emphasised. She was, however, unreceptive to prosecution being described as a tool of last resort. Prosecution, she said, was a 'strategic response': an enforcement tool to be used in appropriate circumstances. To describe it as a tool of last resort would mean that some firms would have no incentive to comply with the laws. These comments are clearly illustrative of the DECC's (NSW) more frequent use of prosecution as an enforcement tool.

D Conclusion

In sum, the DEWHA (federal), the EPA (Victoria) and the DECC (NSW) rarely prosecute polluting offenders. Whilst they see prosecution as an important tool in

44 Statistics taken from EPA Victoria Annual Reports.

45 The remainder of this paragraph is based on discussions with DECC and EPA (Victoria) officials in May 2004.

46 S II B.

47 See ch 3 s II B.

bringing regulated entities into compliance, they are far more likely to use formal administrative enforcement tools and informal techniques such as education and persuasion. Saying that, the EPA (Victoria) and the DECC (NSW) appear to prosecute far more frequently than other state environmental regulators in Australia. But to describe their strategies as 'deterrence-based' would be misleading. In all three cases, a publically available enforcement policy guides regulatory officials in how to exercise their discretionary powers of enforcement. Common factors to take into account in determining the most appropriate course of action in any given case include the seriousness of the offence, the history of the offender and the risk of actual or potential harm.

III Canada

A Introduction

In this section, I examine the enforcement philosophies of regulators in Canada focusing on Environment Canada (federal) and at the provincial level, the Ministry of Environment (Ontario). Traditionally, environmental regulators in Canada have adopted an essentially compliance-based approach to enforcement, especially in the 30 years from 1960 to 1990 when:

> Canada developed all the elements of a compliance system of environmental policy implementation in which recourse to penalties and coercion [existed] only as a very infrequently used last resort of administrators.[48]

But is this still the case?

B Environment Canada

As discussed in Chapter 4 Section III, Environment Canada has an important role to play in enforcing federal environmental law, with the Canadian Environmental Protection Act (CEPA) 1999, which includes pollution prevention provisions such as controlling toxic substances and managing wastes, being one of the most important environmental statutes. But, for many years Environment Canada's enforcement record was viewed as weak. In the mid to late 1990s, Environment Canada's seemingly lax enforcement of federal environmental laws attracted much criticism. One national newspaper, in reporting a 78 per cent drop in prosecution rates between 1992 and 1999, described Canada as the 'promised land' for

[48] M Howlett, 'Policy Institutions and Implementation Styles: The evolution of Instrument Choice in Canadian Environmental Policy' in DL VanNignatten, and R Boardman, (eds), *Canadian Environmental Policy: Context and Cases*, 2nd edn (Toronto, OUP, 2002) 35.

polluters.[49] Evidence presented to the House of Commons Standing Committee on Environment and Sustainable Development in 1998 highlighted a reduction in enforcement action in the mid 1990s, with Dick Martin of the Canadian Labour Congress stating that '[C]learly, Environment Canada is having less and less of an enforcement presence'.[50] The Committee concluded that:

> Whatever the underlying reason or reasons for the inaction, the fact remains that Environment Canada and indeed some provinces are not enforcing environmental laws when they could and should. This failure to act is of deep concern.[51]

The Committee also found serious flaws in the enforcement record under bilateral agreements (whereby the enforcement responsibilities of Environment Canada are delegated to provincial governments) and cited a startling example relating to the Canada-Quebec agreement on the application, in Quebec, of the federal pulp and paper mill regulations:

> [The Sierra Legal Defence Fund] indicated to the Committee that in 1996, at least 20 pulp and paper mills in Quebec had been discharging toxic effluent in excess of the regulated standards, for a total of 189 violations, 98 of which occurred in one mill. It pointed out, however that no prosecutions had been instituted against the offender, thus leading it to conclude that there had been a 'widespread and alarming lack of enforcement by the federal and provincial governments'.[52]

But has Environment Canada's enforcement record improved, and what approach does it currently take to non-compliant activities? Environment Canada's CEPA Compliance and Enforcement Policy is perhaps the best place to start as it provides an interesting insight into the enforcement approach of the federal regulator.[53] The Policy, issued in March 2001, covers 36 pages and gives an extremely detailed breakdown of how, when, why and what compliance and/or enforcement action should be taken in the event of a breach.[54] It is clear from the outset (and indeed is apparent from the Policy's title) that Environment Canada's objective is to secure compliance with CEPA through the use of both compliance promotion activities (such as information, education, communication and consultation) and enforcement mechanisms (inspection, investigation and, in the case of substantiated violations, an enforcement response). Although compliance

[49] DR Boyd, *Unnatural Law: Rethinking Canadian Environmental Law and Policy* (Vancouver, UBC Press, 2003) 238.

[50] Third Report of the Standing Committee on Environment and Sustainable Development *Enforcing Canada's Pollution Laws: The Public Interest Must Come First!* (May 1998) para 49.

[51] *Ibid* para 114.

[52] *Ibid* para 110.

[53] Environment Canada, *Compliance and Enforcement Policy for the Canadian Environmental Protection Act, 1999* (March 2001) available at: <http://www.ec.gc.ca/ele-ale/462A94EB-05D4-4299-93FE-5DBDA127C606/CandEpolicy.pdf> accessed 5 January 2009.

[54] The policy as a whole is informed by five general guiding principles: compliance is mandatory; enforcement must be fair, predictable and consistent; the Act must be administered with an emphasis on prevention of environmental damage; every suspected violation must be examined by enforcement officers and any action taken must be consistent with the Policy; and enforcement officers will encourage regulated entities and others to report suspected violations of the Act. *Ibid* 5.

promotion activities designed to help those subject to CEPA understand and achieve compliance are seen as 'an effective tool in securing conformity with the law'[55], I focus on how enforcement officers designated under CEPA should assess the most appropriate response to an alleged violation as this in more in line with my definition of enforcement in Chapter 1.

In determining the most suitable response to an alleged violation, the Policy stipulates that three factors must be taken in to account when deciding what enforcement action to take.[56] First, enforcement officers must have regard to the nature of the alleged breach including the seriousness of the harm, the intent of the alleged violator and whether it is a repeat occurrence. Second, they must consider 'the effectiveness in achieving the desired result with the violator', the 'desired result' being compliance with the Act, within the shortest possible time and with no further occurrence of violation.[57] Third, it is important to ensure consistency in enforcement. The key factor is the need to secure compliance as quickly as possible, with no recurrence of violation.[58] On this basis, enforcement officers should first consider those responses at the base of the enforcement hierarchy such as warnings and directions which do not require the initiation of formal court proceedings.

The Policy promotes a hierarchy of enforcement tools which can be utilised by either enforcement officers or the Minister in response to a violation. Warnings and tickets can be found at the base of the pyramid and prosecution appears at its apex. The policy favours the use of less formal administrative enforcement measures unless the circumstances of the violation warrant the bringing of civil or criminal action. So, for example, where there is a violation of the Act and the degree of harm or potential harm appears to be minimal, the Policy advocates the use of administrative warnings.[59] Or where there is or is likely to be a release of a substance in contravention of CEPA and its regulations,[60] officials can give 'directions' which require the violator to take all reasonable measures to prevent the release (if it hasn't already occurred), remedy any dangerous condition or reduce any danger to the environment or human health.[61]

The final two enforcement responses at the pinnacle of the enforcement hierarchy are judicial remedies. In certain circumstances, the Minister may seek an injunction in order to stop or prevent a violation.[62] The Minister may also attempt to recover costs by civil suit when, for example, the enforcement officer carried out

[55] *Ibid* 13.

[56] *Ibid* 21.

[57] In this respect, the officer should consider the violator's history of compliance, willingness to co-operate with officers and evidence of corrective action already taken.

[58] *Ibid* 22.

[59] *Ibid* 22–3. Warnings are appropriate where the offender has a good history of compliance and the violator has made reasonable efforts to remedy or mitigate the consequences of the alleged offence.

[60] *Ibid* 23–4.

[61] For certain offences involving the unlawful manufacture, importation etc of a product containing a banned substance, the Minister of Environment can issue a Ministerial Order.

[62] *Ibid* 28–9.

clean-up of a site or took measures to prevent the unauthorised release of a substance.[63] Finally, enforcement officers will lay criminal charges unless a less draconian mechanism is more appropriate. Prosecution will always be pursued in certain circumstances including where there is death of or bodily harm to a person, serious harm or risk to the environment, human life or health or where the violator did not take all reasonable measures to comply with directions or orders. Unusually, the Policy makes no reference to evidentiary criteria which may perhaps explain why in some years the number of convictions is generally quite low compared with the prosecution rate (see Table II below). So, the enforcement policy would suggest that enforcement officers should, except in more serious cases, utilise a response at the base of the pyramid. Reference to the use of persuasive tactics is however noticeably absent from the Policy although this does not mean to say that persuasion does not, unofficially, play an important role in securing compliance and deterring future non-compliance.

A perusal through CEPA annual reports since 1999 would indicate that Environment Canada is increasingly using its formal enforcement powers against violators. However, as illustrated in Table II below, prosecutions (although they fluctuate from one year to the next) are generally low. Environment Canada places much greater reliance on written/oral warnings and directions.[64]

Without access to information on the number of breaches of the legislation (and their seriousness), it is difficult to reach any firm views on Environment Canada's enforcement approach in practice. However, one could probably con-

Table II—Environment Canada Enforcement Statistics—CEPA 1999[65]

	2000/01	2001/02	2002/03	2003/04	2004/05
Investigations	14	56	36	32	43
Warnings	327	513	347	672	1162
Directions	22	5	3	8	2
Environmental Protection Compliance Orders	–	–	1	0	100*
Prosecutions	3	14	4	8	13
Convictions	1	5	3	14	1

* Eighty eight of these were served on dry cleaners for breach of the Tetrachloroethylene (Use in Dry Cleaning and Reporting Requirements) Regulations, compliance with which was designated as a national priority

[63] This would only be necessary where: (1) there was no prosecution; or (2) there was a prosecution but an order for recovery of such costs was not obtained; or (3) where prosecution did not result in conviction. *Ibid* 35–6.

[64] Directions are issued by enforcement officers to deal with or prevent the illegal release of regulated substances into the environment. See further ch 8 s III A iii.

[65] Information taken from annually produced National Statistics for CEPA, available at <http://www.ec.gc.ca/ele-ale/default.asp?lang=En&n=5C63F879-1> accessed 5 January 2009.

clude, albeit tentatively, that the regulator does not adopt a strict deterrence-based approach to enforcement, and will use its powers of prosecution only in limited circumstances. This is reinforced by an OECD Performance Review completed in 2004 which stated that 'litigation has played a relatively small role, particularly in comparison with the US . . . the number of prosecutions remains very small'.[66] The low number of prosecutions, and perhaps more importantly the low conviction rate, could also be indicative of a weak deterrence threat which, as illustrated in Chapter 3, can undermine the effectiveness of more responsive and persuasive approaches to enforcement.[67] Indeed, Environment Canada's enforcement efforts are still subjected to widespread criticism from the environmental lobby, the OECD and other independent review bodies. In discussions with both the Sierra Legal Defence Fund (now Ecojustice) and the Canadian Environmental Law Association (CELA) there was a strong sense of dissatisfaction with Environment Canada's enforcement record.[68]

C The Ministry of Environment (Ontario)

As with Environment Canada, the best place to start in examining the enforcement approach of the Ministry of Environment in Ontario is to consider its Compliance Policy: Applying Enforcement and Abatement Tools, a new version of which was issued in May 2007.[69] The policy is underpinned by the need for provincial officers, at all times to 'seek to work co-operatively and in a professional manner with the responsible person(s) to help address the impacts of a violation and to prevent its recurrence.'[70] In the event of non-compliance, officers can take abatement and/or enforcement action. Terminology is important here. According to the policy, abatement action includes education and outreach (voluntary abatement) and warnings and the issuance of orders (mandatory abatement). An important voluntary abatement tool is the use of an abatement plan which should be developed within a specific period to correct a violation of implement preventive measures. It is undertaken voluntarily, and it is not an offence if the person or persons to whom it applies does not implement the plan's provisions.[71] Other mandatory abatement tools (which are more commonly referred to as enforcement tools elsewhere) include environmental penalty orders (administrative fines) and the

[66] OECD, *Environmental Performance Reviews: Canada* (2004) 122.

[67] See in particular ch 3 s III B, which discusses Ayres and Braithwaite's notion of responsive regulation. They argue that co-operation is best predicted by access to both a range and hierarchy of sanctions but also the punitiveness of the most severe sanction. Where a regulator is extremely reluctant to use a tool such as prosecution, this arguably limits the extent to which a regulated entity will co-operate with a regulator.

[68] Discussions took place in October 2004.

[69] Ministry of Environment (Ontario), *Compliance Policy: Applying Enforcement and Abatement Tools* (May 2007) available at <http://www.ene.gov.on.ca/en/about/penalties/CompliancePolicy.pdf> accessed 5 January 2009.

[70] *Ibid* 1.

[71] *Ibid* 16.

suspension/revocation of authorised documents (although this is classed as an abatement tool of 'last resort).[72] Enforcement action is defined as prosecuting alleged offenders for the purpose of punishing wrongdoing and deterring further non-compliance, including the issuance of tickets or summons. The policy states that responses to violations must be proportionate to the risk presented by the incident, the compliance history of the violator and the response of the violator to the incident.[73]

It is clear from the policy that some incidents warrant the immediate serving of an order or the initiation of an investigation, whilst others may be more suited to a voluntary abatement response. However, if a particular tool does not move the violator into compliance, the Ministry will reassess its approach and may escalate its response accordingly.[74] The policy provides a detailed, documented account of how provincial officers should use an Informed Judgement Matrix (IJM) to evaluate incidents on a case-by-case basis and determine the most appropriate abatement and/or enforcement response. This IJM classifies the severity of the incident and then uses a series of case-specific considerations to determine the recommended response. Stage 1 of the IJM requires officers to determine whether the incident is a violation or has potential for adverse health/environmental impacts.[75] Stage 2 of the process is concerned with evaluating the incident and in doing so, officers must consider the health and environmental consequences, the compliance history of the offender and case-specific considerations such as did the violator voluntarily notify the Ministry of the violation, did they co-operate and what resources did the violator spend on tackling the incident. Stage 3 requires officers to determine the compliance category (see Figure IX below) and then the appropriate enforcement response. Applying the IJM categories will not, however, lead to conclusive evidence that an incident should be placed in a particular compliance category. The policy stresses the need for officers to make an informed judgement and there may be additional factors that would move the incident from one category to another.[76] Finally, Stage 4 relates to monitoring and following-up the compliance status of the violator.

Prosecutions, the primary 'enforcement' response, are an important part of the Ministry's enforcement approach and should be available to deter serious violations.[77] All three compliance categories permit officers to refer cases to the Investigations and Enforcement Branch of the Ministry, who will investigate suspected violations in order to determine whether there are reasonable and prob-

[72] Ministry of Environment (Ontario), *Compliance Policy: Applying Enforcement and Abatement Tools* (May 2007) available at <http://www.ene.gov.on.ca/en/about/penalties/CompliancePolicy.pdf> accessed 5 January 2009, 18–19.

[73] *Ibid* 4.

[74] *Ibid.*

[75] This is important as even if an incident does not constitute a violation of regulatory controls, officers may have other grounds for issuing a control document.

[76] *Ibid* 12–13.

[77] Taken from Ministry of Environment Enforcement Policy, *Ibid* 13.

Compliance Category I recommends tools that encourage the responsible person to attain compliance voluntarily. The best course of action may be education and outreach to help the responsible person understand its regulatory environmental obligations. However, orders, EP Orders, tickets and IEB referrals may also be considered in Compliance Category I.

Compliance Category II recommends stronger mandatory application of tools such as orders, EP Orders, and use of POA tickets. An IEB Referral shall be considered for incidents that fall in Category II, except when a POA ticket is used.

Compliance Category III recommends the use of a mandatory abatement tool such as the issuance of a control document, to resolve the incident. Incidents that fall in Category III must also be referred to IEB

** IEB—Investigations and Enforcement Branch

Figure IX: Ministry of Environment (Ontario) Compliance Policy—Compliance Categories[78]

able grounds for the laying of criminal charges.[79] In doing so, they are required to have regard to a range of factors including the severity of the violation, the mens rea of the offender, the offender's compliance history, the extent to which the offender co-operated with the regulator and the deterrent effect of enforcement action on similar or other operations.[80] Consideration is also given, at this stage, to whether there is a reasonable prospect of conviction and whether prosecution would be in the public interest.[81]

All in all, the Ministry's policy suggests that provincial officers should, where appropriate, adopt a compliance-based enforcement approach, with an emphasis on the use of voluntary measures and abatement tools in all but the most serious cases of non-compliance. The policy also has the feel of a quite responsive approach to enforcement, with its emphasis on moving up the enforcement ladder if a particular instrument has not moved a regulated entity into compliance.

But does enforcement on the ground mirror the principles the Ministry advocates in its most recent compliance policy? Over the past 30 to 35 years the enforcement approach of the Ministry has varied considerably. According to Dianne Saxe, in the 1970s and early 1980s, 'environmental prosecutions received very little emphasis . . . regulators concentrated on administrative remedies almost to the exclusion of prosecution.'[82] However, in the early 1980s, in response to public calls

[78] *Ibid* 12.
[79] *Ibid* 21.
[80] *Ibid.*
[81] *Ibid* 21.
[82] D Saxe, *Environmental Offences: Corporate Responsibility and Executive Liability* (Toronto, Canada Law Book, 1990) 29–30.

for greater compliance, judicial demands that enforcement be improved, and changing attitudes within the MoE, the government began to revise its approach. It established a Special Investigations Unit and in 1985 created an Investigations and Enforcement Branch (IEB) of the Ministry. With these changes, the 'MoE became increasingly enforcement orientated and the number of prosecutions increased dramatically.'[83] Between 1985 and 1995, staff numbers within the IEB increased, the liability of company officers and directors was expanded, higher fine levels were introduced into Ontario's environmental laws and provincial officers were given new investigating powers. The Canadian Institute for Environmental Law and Policy (CIELAP) has described the years 1985–95 as 'a period of relative prosperity in terms of environmental compliance and enforcement initiatives.'[84] But in 1995, when the Progressive Conservative Party came into power, the province witnessed a sea change in compliance philosophy.[85] A CIELAP report published in 1999 concluded that during 1995–99, there was 'a precipitous decline in the province's environmental law enforcement activities' with total fines at their lowest level since 1986/87.[86]

Criticism continued into 2000, when a report produced by Ontario's Provincial Auditor stated that there were many instances where environmental officers 'did not follow up on violations to ensure that the facility operator had subsequently corrected the deficiency and responded appropriately.'[87] In fact, an internal review carried out by the Ministry during that time determined that in 69 of the 100 inspection reports reviewed, a number of violations were identified, including 22 considered to be significant. In terms of enforcement action, only one control order was issued and no fines or charges were laid. In 19 cases, the officer requested that the operator provide a voluntary abatement action plan but only one plan was ever received.[88] The events that occurred in Walkerton in May 2000 also had a profound impact on the Ministry's attitude towards enforcement. A public inquiry was launched to investigate how the municipal drinking water system in Walkerton had become contaminated with deadly bacteria, killing seven people and making 2,300 people ill. The long-term environmental impact of the incident is still being assessed.[89]

In response to the concerns expressed by the Provincial Auditor, CIELAP and others, in 2000 the Ministry revisited its approach to enforcement and compliance and there is some evidence to suggest that there is a corresponding increase in formal enforcement activity. Before examining this, there are two important matters

[83] CIELAP, *Sixth Annual Report on Ontario's Environment* (January 2002) 73.

[84] *Ibid* 72.

[85] See further ch 4 s II B ii for details of budget and staffing cuts.

[86] CIELAP, *Ontario's Environment and the Common Sense Revolution—A Four Year Report* (1999) 7 available at: <http://www.cielap.org/pdf/csr4.pdf> accessed 5 January 2009.

[87] Officer of the Auditor General of Ontario, *Ministry of Environment Operations Division: Special Report* (2000) 120.

[88] *Ibid.*

[89] For details of the inquiry following the incident, see <http://www.attorneygeneral.jus.gov.on.ca/english/about/pubs/walkerton/> accessed 5 January 2009.

to bear in mind. First, the Compliance Policy discussed above post-dates many of the enforcement statistics referred to below. However, its principles and approaches are very similar to those advocated in its previous policy of October 2001, although the Informed Judgement Matrix is a new initiative. Second, primary statistical data on enforcement action in Ontario is extremely hard to come by. During the 1990s, the Ministry stopped publicising its enforcement statistics although groups such as Sierra Legal Defence Fund (now Ecojustice) have used the Freedom of Information Act to compile several reports on enforcement practices. With the exception of Table III below, the enforcement statistics that follow have been taken from these external sources. Table III summarises prosecutions brought by the Ministry from 2001 to 2003.

Table III—Prosecution Statistics 2001–2003[90]

	Nov 01–Nov 02	Nov 02–Nov 03
Prosecutions commenced	146	178
Prosecutions not proceeded with	10	21
Conviction after trial	12	18
Guilty Plea	88	138
Acquittal	4	10
Withdrawal	9	9

On the face of it, there seems to be an increase in the number of prosecutions. However, they are still relatively rare, with commentators highlighting the current emphasis on negotiation and conciliation:

Warnings and extra time for compliance may be substituted for formal investigations and prosecution. Technical violations may be overlooked in return for agreements to invest resources into technologies or practices that address larger problems. Even clear violations may be tolerated if the violation is not serious or it appears that the regulated person is genuinely attempting to be compliant.[91]

An influential report produced by the Sierra Legal Defence Fund (now Ecojustice) examined the extent of non-compliance with Ontario's air and water pollution laws.[92] It revealed that compared with the number of violations, enforcement action in terms of prosecution is very low. For example, in 2001 there were more than 2,300 violations of provincial wastewater laws including 1,952 violations of wastewater limits, over 200 reported spills and sewage bypasses and 152 administrative violations.[93] In response, the MoE launched 20 investigations

[90] Information provided by the Ministry of the Environment (Ontario) (October 2004).

[91] S Verhulst, 'Legislating a Principled Approach to Sentencing in Relation to Regulatory Offences' (2008) 12 *Canadian Criminal Law Review* 281, 285.

[92] Sierra Legal Defence Fund, *Cracking Down on Polluters: A Report on Violations of Ontario's Air and Water Pollution Laws* (March 2004).

[93] *Ibid* 3.

pertaining to non-compliant water discharges. Of these, only six companies were charged at the time the documents were received. 10 of the investigations were dropped due to either insufficient evidence or voluntary abatement measures, three were still under investigation and one investigation was closed for unknown reasons.[94]

Prosecutions also appear to account for a very small proportion of the Ministry's enforcement activity, with much greater store being placed on less draconian enforcement and abatement mechanisms. For example, in 2002, the Auditor General revisited the issue of environmental enforcement and found that enforcement activity appeared to be on the increase, reporting that the number of control orders issued requiring corrective action had risen from less than 500 in 1999/2000 to about 1,500 in 2000/01 and approximately 2,200 in 2001/02. Compliance reports for air emissions, municipal sewage discharges and industrial sewage discharges in 2005 also reveals that where abatement and enforcement action is taken there is an almost overwhelming reliance on provincial officers' orders and voluntary abatement programmes. In a large number of cases, incidents were resolved and action was taken without recourse to any formal abatement or enforcement mechanism.[95]

So, all in all, despite recent improvements in enforcement activity, there is still widespread dissatisfaction with the Ministry's enforcement record. For example, an Ecojustice media release of June 2005 (applauding the introduction of environmental penalty orders (administrative monetary fines)) comments that:

> [C]orporate polluters have had little to fear up until now. Prosecutions have been rare, drawn out, and often result in slight reprimands, miniscule monetary fines or outright dismissals for technical reasons.[96]

Also, in its report *Doing Less with Less* the Environmental Commissioner for Ontario found that where cases of non-compliance are identified in sewage treatment plants, the Ministry prefers to apply 'softer, unenforceable approaches (such as voluntary abatement programs)'.[97] The Commissioner reported that in 2004–05 the Ministry issued formal enforcement orders to only eight per cent of municipal facilities who were found to be in breach of the regulations. The remainder were dealt with by voluntary measures. This, the Commissioner concluded, 'may send the misleading message that deficiencies in STPs [sewage treatment plants] are not a serious concern.'[98] There also appears to be a particular problem in tackling persistent polluters. According to a Sierra Legal Defence Fund

[94] *Ibid* 8.

[95] Ministry of Environment, 2005 Environmental Compliance Reports available at <http://www.ene.gov.on.ca/envision/compliance/comp05/index.htm#graphs> accessed 5 January 2009.

[96] Ecojustice, 'Ontario Passes Polluter Pays Bill' (19 June 2005) available at <http://www.ecojustice.ca/media-centre/press-releases/ontario-passes-polluter-pays-bill> accessed 5 January 2009. For more detail on environmental penalty orders in Ontario see ch 8 s III A iii.

[97] Office of the Environmental Commissioner of Ontario, *Doing Less with Less: How Shortfalls in budgets, staff and in-house expertise are hampering the effectiveness of the MOE and MNR* (April 2007) 32.

[98] *Ibid.*

report published in 2004, the top six wastewater pollution violators for 2001 committed a total of 920 violations of wastewater regulation.[99] All six companies were non-compliant for at least four years and yet none of them was prosecuted. The only enforcement action involved issuing control or provincial officer's orders, or establishing a voluntary abatement programme. The report concludes that:

> Prosecution of the top violators would result in a large improvement in non-compliant numbers since a handful of facilities are responsible for a large percentage of the violations.[100]

The Provincial Auditor and groups such as CIELAP have echoed similar sentiments, urging the government to tackle the problem. Since the publication of many of these reports, the Ontario government has passed legislation which gives Ministry officials the power to issue environmental monetary penalties when industrial facilities spill hazardous chemicals. This will be an important weapon in the Ministry's armoury for tackling industrial pollution, and will probably lead to a drop in prosecution numbers.[101]

D Conclusion

Environment Canada and the Ministry of Environment (Ontario) can both use a range of enforcement tools in pursuing regulatory compliance. Breach of pollution control regulation attracts criminal liability and prosecution is seen as an important part of the enforcement strategy. However, prosecution is undoubtedly the exception rather than the rule and mandatory and voluntary abatement techniques such as warnings and orders are a more likely response to regulatory violations. Both regulators have produced and published guidance on enforcement. The compliance policy of the Ministry of Environment (Ontario) is particularly comprehensive, prescribing an Informed Judgement Matrix which is used to classify incidents, and a staged process which must be followed in determining appropriate action. Overall, enforcement activity would certainly seem to have increased since the budgets cuts of the late 1990s. Nevertheless, there still seems to be a sense of dissatisfaction with the enforcement record of both the Ministry and Environment Canada. The deterrence threat of enforcement action is perhaps not as strong as it could be.

[99] Above n 92.
[100] *Ibid* 8.
[101] These penalties will be discussed in more detail in ch 8 s II D.

IV England and Wales

A The Enforcement Approach of the Environment Agency

As evidenced in empirical studies of environmental enforcement activities in the UK in the 1980s,[102] environmental agencies adopted a predominantly compliance-based approach.[103] Regulatory officials relied on informal enforcement and compliance mechanisms such as persuasion and verbal warnings and in doing so, emphasised the importance of maintaining a good relationship with regulated entities. Prosecution was regarded 'not only as reflecting badly on the company affected but also as a failure on the part of the regulatory officials.'[104] However, it was also evident that where more persuasive tactics failed, officers would use increasingly formal and punitive methods such as administrative notices, licence suspension and criminal prosecution. As summarised by Hawkins:

> The officer has, in practice an array of moves at his disposal in seeking compliance. The tactics are employed serially, the sequence moving from a more conciliatory to a more coercive approach, if conciliation fails.[105]

The Environment Agency's approach to enforcement today is arguably more formal, deterrence-based and legalistic than that of its predecessors. Saying that, prosecutions are still very rare and a large number of breaches of environmental law each year are not enforced in any formal sense.

Before examining enforcement statistics, a valuable insight into the approach of the Environment Agency can be gained from looking at its enforcement policy. The Agency's Enforcement and Prosecution Policy is underpinned by four principles of enforcement: proportionality in applying the law and securing compliance, consistency of approach, transparency about how the Agency operates and what the regulated community can expect from the Agency and the targeting of enforcement action.[106] The purpose of enforcement is to 'ensure that preventative or remedial action is taken to protect the environment or to secure compliance with a regulatory system.'[107] It highlights the importance of a co-operative approach and proposes to achieve a better environment through education and advice.[108] In doing so, '[the Agency] offers information and advice to those it regulates and seeks to secure co-

[102] See, eg, G Richardson, AI Ogus and P Burrows, *Policing Pollution: A Study of Regulation and Enforcement* (Oxford OUP, 1983); K Hawkins, *Environment and Enforcement: Regulation and the Social Definition of Pollution* (Oxford, OUP, 1984).

[103] See generally D Vogel, *National Styles of Regulation: Environmental Policy in Great Britain and the United States* (London, Cornell University Press, 1986) ch 2.

[104] *Ibid* 87.

[105] Above n 102 (Hawkins), 130.

[106] Environment Agency, Enforcement and Prosecution Policy (August 2008) paras 9–19, available at <http://www.environment-agency.gov.uk/static/documents/enforcement-policy.pdf> accessed 7 January 2009.

[107] *Ibid* para 6.

[108] *Ibid* para 1.

operation avoiding bureaucracy and excessive cost.'[109] Whilst there is an emphasis on encouraging and facilitating compliance, the policy states that other enforcement tools, including prosecution, will be used where the situation warrants such action. It is made clear that the Agency will not hesitate to use its enforcement powers where appropriate and if a criminal offence is committed, 'in addition to any other enforcement action, the Agency will consider instituting a prosecution, administering a caution or issuing a warning.'[110] In deciding whether or not to prosecute a suspected offender, the purpose of which is 'to punish wrongdoing, to avoid a recurrence and to act as a deterrent to others', Agency staff are required to take into account a number of public interest factors including the intent, attitude and history of the offender, the environmental effect of the offence and its foreseeability.[111] Even if a prosecution would seem to be in the public interest, there is a strict evidential test in that the Agency will only commence a prosecution if it is satisfied that 'there is sufficient, admissible and reliable evidence that the offence has been committed and that there is a realistic prospect of conviction.'[112] In some cases, there is a presumption in favour of prosecution. Where there is sufficient evidence, the Agency will normally prosecute where, for example, an incident or breach will have significant consequences for the environment, breaches are excessive or persistent or there has been a failure to comply adequately with formal remedial requirements such as those specified in an enforcement notice.[113]

The Agency's Enforcement and Prosecution Policy should be read in conjunction with the recently written Guidance to the Enforcement and Prosecution Policy.[114] These guidelines indicate the normal response that would be taken for specific offences relating to, for example, water quality, waste, IPC/IPPC and water resources. In determining this 'normal' enforcement response, the guidelines refer to two different 'types' of mechanisms: preventative/remedial actions such as enforcement notices, suspension/revocation of licences, injunctions and Anti-Social Behavioural Orders and criminal enforcement responses such as prosecutions, formal cautions and warnings. Whilst the guidelines stipulate a normal response for specific offences, they also require enforcement officers to take into account public interest factors such as the nature of the offence, the deterrent effect of enforcement and the intent of the offender.[115] Another important factor is the environmental effect of the offence. This is assessed using the Agency's Common Incident Classification Scheme (CICS) which categorises offences according to their actual (or potential) environmental impact: 'major' (Category

[109] *Ibid* para 4.
[110] *Ibid* para 8.
[111] *Ibid* paras 20 and 23.
[112] *Ibid* para 22. The evidential and public interest tests are reflected in the more general Code for Crown Prosecutors.
[113] *Ibid* para 29.
[114] Environment Agency, Guidance for the enforcement and prosecution policy: operational instruction (November 2008) available at: <http://www.environment-agency.gov.uk/static/documents/epp-guidance.pdf> accessed 7 January 2009.
[115] *Ibid* para 1.5.

1), 'significant' (Category 2) or 'minor' (Category 3). There is a fourth category to cover incidents where there is no environmental impact. In addition, the Agency has introduced a new Compliance Classification Scheme (CCS) which classifies non-compliance with permit conditions according to their severity. There is no need for actual environmental harm under the CCS, but where breach of a condition leads to actual environmental impact, there is clearly an overlap between the CCS and the CICS.[116]

All in all, the Agency's enforcement policies and guidelines envisage the use of a range of enforcement mechanisms, from more informal, persuasive techniques to formal criminal, civil and administrative sanctions. It is clear throughout that where appropriate, the Agency will not hesitate to use its powers of prosecution. But to what extent does practice on the ground actually mirror the Agency's policy approach? In the overwhelming majority of cases of non-compliance, the Agency does not instigate prosecution proceedings. For example, in 1999–2000, 17,592 waste incidents were investigated by the Agency, but only 342 offenders were prosecuted.[117] And as regards water pollution offences under the Water Resources Act 1991, 246 prosecutions resulted from the investigation of 14,417 'substantiated' pollution incidents.[118]According to Bell and McGillivray, there are around 25,000 plus pollution incidents per year[119] but, as can be seen from Figure X below, only a relatively small number of prosecutions are brought. In fact, the Agency prosecutes less than five per cent of breaches.

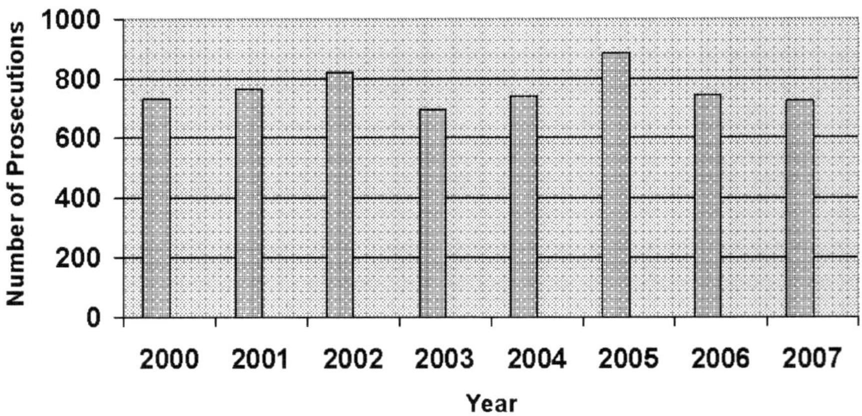

Figure X: Environment Agency Prosecutions 2000–2007[120]

[116] For additional information, see Environment Agency, *Compliance Classification Scheme Generic Guidance* available at <http://www.environment-agency.gov.uk/static/documents/Business/generic_guidance_1195099.pdf> accessed 7 January 2009.

[117] Environment Agency Annual Report and Accounts 1999–2000, 131.

[118] *Ibid* 129.

[119] S Bell and D McGillivray, *Environmental Law*, 6th edn (Oxford, OUP, 2006) 295–6.

[120] The data relating to 2000–04 is taken from ENDS Report, 'Agency to target "big, bad and nasty" crime' (2005) 364 ENDS Report 14–15. Years 2005–07 are taken from the House of Lords Written Answers (Hansard) Col WA193 (6 February 2008).

However, as indicated in the Enforcement and Prosecution Policy, prosecution is seen as an important tool in the Agency's enforcement weaponry with a recent report stating that '[P]rosecutions are our last resort, but we will prosecute if we have to.'[121] In fact, statistics would strongly suggest that where the Agency sees fit, offences will be prosecuted. For example, one could compare the number of prosecutions with the number of Category 1 (major) incidents; that is incidents which are most likely to generate a response in the form of a prosecution (see Table IV below).

Table IV—Prosecutions and Category 1 Incidents[122]

	Category 1 Incidents	Prosecutions
2000	99	730
2001	147	766
2002	111	820
2003	111	693
2004	131	740
2005	108	887
2006	92	744

However, in contrast with the environmental regulators in Canada and Australia, discussed earlier in this chapter, prosecution numbers are high compared with the use of other formal enforcement powers. As illustrated in Table V below, from 2000–04, the number of cautions and administrative notices issued was lower than the number of prosecutions.

Table V—Environment Agency Enforcement Action 2000–2004[123]

	Prosecutions	Cautions	Notices
2000	733	353	389
2001	786	334	434
2002	820	408	371
2003	693	338	396
2004	740	345	563

[121] Environment Agency, *Spotlight in Business: Environmental Performance in 2006* (July 2007) 10.

[122] Statistics on the number of Category 1 incidents are provided by the Environment Agency and are available at <http://www.environment-agency.gov.uk/research/library/data/34363.aspx> accessed 7 January 2009

[123] Taken from ENDS Report above n 120.

B Conclusion

In summary, it would seem that the majority of breaches of environmental law in England and Wales are not dealt with through the use of formal enforcement action such as cautions, enforcement notices or prosecutions. This would suggest that in most cases, an informal, compliance-based approach is preferred in tackling environmental offenders. Like regulators in Australia and Canada, the Agency has produced an enforcement policy which outlines the factors to be taken into account in determining the most appropriate enforcement action.

The Agency's enforcement approach would, however, appear to be significantly different from that of its counterparts in Australia and Canada. Most notably, it utilises criminal prosecution more than it does other formal enforcement tools such as cautions and enforcement notices. As illustrated in Sections II and III above, Australian and Canadian environmental regulators are less likely to use prosecution than other formal enforcement tools.

V Conclusion

Whilst it is very difficult to: (a) classify an enforcement approach without having access to inter alia data on the number of pollution 'incidents' and their seriousness; and (b), in the light of this, make any valuable comparisons in enforcement between different jurisdictions, this chapter has provided an insight into pollution control law enforcement in Australia, Canada and England and Wales. Formal enforcement responses such as criminal prosecution and administrative notices are important in pursuing regulatory compliance. They are, however, seen more as a 'last resort' and often signify a breakdown in communication between the two parties. This is not to suggest that the regulators never utilise formal enforcement tools. As illustrated in previous sections, regulators do issue warnings, do serve administrative enforcement notices and do prosecute for environmental offences. But the use of such instruments is rare and in most cases, the regulators appear to resort to more informal methods in order to coax a regulated entity into compliance.

With the exception of the Environment Agency in England and Wales, regulators are more likely to use formal enforcement mechanisms such as administrative notices and monetary administrative penalties than criminal prosecution. Nevertheless, all the enforcement policies stress the importance of prosecution in deterring environmental crime. It is seen as a vital weapon in the enforcement battle, with guidance stipulating factors that should be taken into account in deciding whether or not to prosecute. The seriousness of the offence, the intention and history of the offender and the extent to which the offender has co-operated with the regulator are just three examples of factors that are relevant in exercising prosecutorial discretion. A hierarchy of enforcement mechanisms underpins (either explicitly or implicitly) the enforcement approaches of the regulators, with regulators determining the most appropriate response in accordance with criteria laid out in their respective enforcement policies. In the case of the Department of Environment, Water, Heritage and the Arts (Australia), explicit reference is made to a responsive enforcement approach which escalates according to the seriousness of the offence of the length of time during which non-compliance activities continue.

Formal enforcement action, particularly criminal prosecution is, of course, a costly exercise for public regulators, especially when many are under-staffed and under-resourced. But compliance-based, persuasive strategies such as those adopted by our environmental regulators, are only effective if there is a credible deterrence threat. This is most obviously achieved through prosecution but other formal tools such as administrative notices and monetary administrative penalties can all deter non-compliance. In Chapters 6, 7 and 8, I examine in detail the deterrent effect of a range of formal enforcement mechanisms, beginning in Chapters 6 and 7 with criminal prosecution and criminal sanctions. Using the theoretical framework developed in Chapters 1 and 2, I analyse the extent to which a range of interesting and innovative enforcement tools can deter offenders and promote cost-effective enforcement.

6

Criminalising Polluting Behaviour: Models of Criminal Liability and Deterrence

I Introduction

IN JULY 1997, Mr John Moynihan, the director of Green Environmental Industries Ltd, was sentenced to 18 months imprisonment for the unlawful storage and deposit of clinical waste including dressings, needles, syringes and human tissue.[1] In 1994, Mr Moynihan, the defendant set up Green Environmental Industries Ltd with a view to incinerating the increasing amount of clinical waste originating from the private health care sector. However, before the company's incineration facility was completed and the necessary licences were obtained, the defendant began to accept large volumes of clinical waste from licensed waste disposal carriers employed by various establishments, including hospitals, veterinary surgeries and clinics. When attempts to have the waste dealt with by other operators failed, the defendant stored it in a number of hired containers and lorry trailers. Mr Moynihan received a 10-month custodial sentence for this offence. When asked by the Waste Regulation Authority[2] to remove and dispose of the waste lawfully, Mr Moynihan proceeded to dump the waste at a disused factory. For this offence, he was given an 18-month custodial sentence to run concurrently with the 10-month sentence previously awarded. Mr Moynihan was also sentenced to nine months for forgery and six months for two offences of dishonesty, to run concurrently within themselves but to run consecutively with the 18-month sentence imposed for the environmental offences. On appeal, the judge determined that all the custodial sentences should run concurrently.[3]

[1] Unreported, St Alban's Crown Court, 25 July 1997.

[2] Prior to the creation of the Environment Agency in 1995, Waste Regulation Authorities were responsible for administering and enforcing waste management provisions under the Environmental Protection Act 1990. For a summary of Environment Agency responsibilities see ch 4 s IV B.

[3] [1999] 1 Cr App R (S) 294.

The initiation of a criminal prosecution is a potentially powerful tool at the hands of environmental regulators and is a common feature of pollution control regulation. In many (if not most) jurisdictions, including Australia, Canada, and England and Wales the public enforcement of pollution control law is underpinned by the imposition of criminal liability. Although it is difficult to generalise, pollution control statutes ordinarily utilise a number of different models of criminal behaviour. One model prohibits all forms of polluting activity, even if a licence is in force and the activities are operated in accordance with that licence. For example, the Ontario Water Resources Act makes it an offence to discharge or cause or permit the discharge of any kind of material in or into any waters where that discharge may impair the quality of the water.[4] A second model prohibits persons from engaging in legislatively defined activities, unless that person is both acting under a licence or permit issued by the regulatory authority and satisfying the conditions attached to this permit.[5] Here, except in cases where emissions limits and operational standards are laid down in legislative acts, the criminal law supports a system of administrative decision-making whereby the regulatory authority establishes environmental standards, and incorporates them into the licence or permit by the use of conditions. As a result, the administrative authority has considerable power in fixing the conditions of criminal liability.[6] For example, under the pollution prevention and control regime in England and Wales, it is an offence 'to operate an installation or mobile plant except under and to the extent authorised by a permit'.[7] Perhaps in response to the potential harshness of imposing criminal responsibility on those who had no knowledge of the act or could not have avoided the act, the strictness of the offences under both models is sometimes mitigated by the availability of statutory defences.[8]

Whatever the model used, advocates of the criminal law emphasise its effectiveness as a deterrent, particularly when applied to companies that arguably fear the stigmatising effect of criminal prosecution more than any criminal sanction that is imposed consequent to conviction. This, the argument goes, cannot be achieved by the taking of civil or indeed administrative enforcement action.

According to the deterrence framework advanced in Chapter 2 (see Figure XI below), the deterrent impact of the criminal law depends on (a) the likelihood of

[4] S 30(1).

[5] This technique has been described as involving a 'legislative definition of unlawful conduct by means of absolute prohibition coupled with a formal delegation of partial exemption'. See G Richardson, A Ogus, and P Burrows, *Policing Pollution: A Study of Regulation and Enforcement* (Oxford, Clarendon Press, 1982) 46.

[6] There are, however, limits to the administrative authority's power in determining the scope of licensing conditions. See further M Faure and M Visser, 'How to Punish Environmental Pollution? Some Reflections on Various Models of Criminalization of Environmental Harm' (1995) 4 *European Journal of Crime, Criminal Law and Criminal Justice* 316, 321.

[7] Pollution Prevention and Control (England and Wales) Regulations 2000 SI 2000/1973 reg 32.

[8] For details of liability requirements and available defences, see s III below.

$$B < gH + eF + cD$$

Using this framework, enforcement will induce desirable behaviour (ie compliance) where *B* (the utility/benefit derived from non-compliance) is less than *gH* (the probability and associated costs of detection by an enforcement agency including the hassle costs associated with detection), *eF* the probability and associated costs of administrative enforcement action (such as the imposition of a monetary administrative penalty or an enforcement notice) and *cD* the probability of formal court action (both civil and criminal) and its associated costs (including a court-based sanction). The *cD* element of the model includes, most obviously, the probability of prosecution and conviction in the criminal courts

Figure XI: The Deterrence Framework

prosecution and conviction, and (b) the scale of the anticipated penalty. Both these components are influenced by a number of factors relating to the nature and scope of pollution control regulation. Procedural and evidential requirements clearly influence the likelihood and outcome of proseuction. Although they vary between jurisdictions, criminal procedures generally include trial by jury for the most serious offences, a higher burden of proof on the prosecution than that used in non-criminal cases, a mens rea requirement (such as intention or negligence) and rules of evidence protecting the defendant. Compared with sanctions imposed through a civil or administrative process, the costs of the criminal justice system are high. This is likely to go some way to explaining the small number of prosecutions relative to polluting incidents. Investigating a pollution incident and gathering enough evidence to prove 'beyond reasonable doubt' the guilt of the defendant has serious resource-implications, especially in environmental cases where securing reliable evidence and establishing causal links between the defendant's acts and actionable behaviour are often difficult. As a result, criminal prosecutions are used selectively where, for example, the case involves serious harm or risk to the environment or where there has been deliberate non-compliance.[9] The principal purpose of these procedures is to reduce the number of errors, such as wrongful convictions, resulting from the adjudication process. But since the procedures designed to reduce these errors are themselves costly, there is, in principle at least, an optimal level of error reduction, approximate to the costs.

One way of reducing the time and cost associated with securing a conviction is by modelling criminal behaviour on the basis of strict as opposed to mens rea or fault-based liability. This negates the need for the prosecution to prove a guilty mind and can thus reduce the costs associated with bringing an action. A regulator

[9] For a summary of prosecution policies in the chosen countries, states and provinces see ch 5.

may therefore be more inclined to take formal enforcement action (whether criminal or indeed civil or administrative) where the offence is strict, thereby raising the likelihood of prosecution in the case of apprehension. The deterrent impact of prosecution, as determined by weighing up the probability of prosecution and its associated costs, against the benefit the offender derives from non-compliance, is also strengthened by the fact that it is easier to secure a conviction for strict liability crimes. So, from an economic perspective at least, strict liability is arguably desirable. But the debate is more complex than this. Strict liability potentially also explains low judicial penalties which, in and of itself, weakens the deterrent impact of the criminal law. Furthermore, not all debates on strict versus fault-based liability tie into the economic deterrence model: the criminal law's focus on notions of moral culpability, fairness and legitimacy can conflict with the economist's view of strict liability as efficient and effective. Section II begins by examining arguments for and against the use of no-fault/strict liability offences in pollution control regulation. A second and related factor which arguably influences the scale of the expected penalty is the perception that regulatory offences (partly by virtue of the fact that they utilise strict liability approaches) cannot be reconciled with traditional (and indeed popular) notions of crime. Critics argue that the judiciary are unwilling to view pollution control offences as 'real crimes'. By directly influencing the expected penalty, the deterrence function of the criminal law is again reduced. Section II concludes, in subsection C, by discussing the real versus regulatory crimes debate. Section III then discusses liability rules in pollution control regulation in Australia, Canada, and England and Wales.

The chapter finishes in Section IV by considering the deterrent impact of corporate and directorial liability and its practical application in Australia, Canada, and England and Wales. Many pollution incidents are caused by companies and the deterrent effect of the criminal law is undoubtedly strengthened where corporate entities can be held liable for pollution control offences. Not only are they responsible for a large percentage of serious pollution incidents but they are also most vulnerable to the adverse publicity associated with prosecution. Company directors and other senior officers can also be held to account by the criminal law. By virtue of the fact that they will probably be responsible for designing corporate strategy and policy, they are arguably most influential in maximising corporate compliance with pollution control regulation. But attributing such responsibility, both corporate and directorial, is difficult, particularly so under the criminal law where the 'beyond reasonable doubt' burden of proof applies. All in all, the difficulties inherent in proving both corporate and directorial liability weaken the potentially significant deterrent impact of pursuing companies and/or corporate executives through the criminal courts. The higher the potential costs of prosecution, the less likely it is that prosecution is a viable enforcement response.

II Deterrence, Rules of Liability and Perceptions of Crime

A Introduction

As noted in the introduction, the criminal justice process has a number of procedural arrangements, all of which impose costs on both the regulator and the regulated. These primary administrative costs undoubtedly influence the volume of criminal prosecutions brought by pollution control regulators, which in turn limits the deterrent impact of the criminal law. There are also, however, a set of secondary costs which arise as a consequence of criminal prosecution, most notably error costs, which are the 'adverse consequences to particular individuals and to society more generally if the processes generate some inappropriate condemnations and sanctions'.[10] The purpose of the procedural arrangements (such as higher burden of proof and mens rea) is to reduce any errors costs that may result from the adjudication process. One category of errors costs is the imposition of inappropriate penalties (referred to as Type I errors). The most obvious benefit here is to the innocent parties themselves who will avoid penalties and other associated losses such as loss of reputation consequent to conviction. But the costs associated with this also need to be taken into account. These may include an increase in administrative resources and the fact that some guilty defendants may avoid a penalty (referred to as Type II errors). The key then, from an economic perspective at least, is to achieve an optimal level of error reduction where the benefits are approximate to the costs.[11] We can examine criminal rules of liability with this in mind.

The scope of criminal pollution control offences is determined, to some extent, by legislatively prescribed rules of liability which stipulate the degree of mens rea required to prove an offence. Although these rules vary, strict liability offences (by this I mean those offences for which a person may be liable in the absence of intention, knowledge, recklessness or negligence)[12] are a common feature of pollution control regulation in many jurisdictions including Australia, Canada, and England and Wales. In all three countries:

[10] A Ogus, M Faure and N Philipsen, *Best Practices for Consumer Policy: Report on the Effectiveness of Enforcement Regimes* (OECD, 2006) para 39.

[11] See RA Posner, 'An Economic Approach to legal Procedure and Judicial Administration' (1973) 2 *Journal of Legal Studies* 399.

[12] There is no commonly accepted definition of strict liability. For the purposes of this section, I will also include as strict liability offences, those which are committed without 'fault' but which can be avoided on proof of certain defences, such as where the defendant acted with 'due diligence' or took 'reasonable precautions'. See further A Ashworth, *Principles of Criminal* Law, 5th edn (Oxford, OUP, 2006) 164–5.

> The removal of the common law requirement for a mental element in 'public welfare' legislation has been justified on the basis of protecting the community by enforcing a high standard of care.[13]

But can strict liability really be justified from an economic (or indeed other) perspective? Section B below examines the merits of strict liability, before moving on in Section C to discuss the real versus regulatory crime debate.

B Strict Liability: Friend or Foe?

From an economic perspective, strict liability is conducive to deterring undesirable behaviour. As noted earlier in this chapter, where prosecution is too costly, the probability of prosecution (and therefore conviction) is lowered ('c' in the deterrence framework) and the deterrent effect of such action is reduced.[14] Proof of mens rea can impose prohibitively high administrative costs whereas strict liability supports enforcement efficiency by 'lowering the cost of regulator investigation, intrusion on business, prosecution and defence effort, and reducing trial length.'[15] Economic considerations would therefore favour strict liability as it eases the procedural burden on enforcing authorities who pursue offenders through the criminal courts. Regulatory authorities are also in a stronger position to use the threat of prosecution and conviction in order to secure compliance.[16] This is of paramount importance when one considers that environmental regulators in Australia, Canada, and England and Wales generally adopt a 'compliance' approach to enforcement, whereby prosecution is used as a last resort and even then, only where the regulatory contravention is significant.[17] In summary, the availability of a no-fault offence strengthens the deterrence function and enhances the ability of the regulator to ensure compliance.

Moving away from economic considerations, strict liability can arguably be justified on the basis that it supports the role of the criminal law in protecting fundamental social interests by easing the burden on prosecuting authorities. As put by Wootton:

> [I]f the object of the criminal law is to prevent the occurrence of socially damaging actions, it would be absurd to turn a blind eye to those which were due to carelessness, negligence or even an accident. The question of motivation is in the first instance irrelevant.[18]

[13] Australian Law Reform Commission, *Principled Regulation: Federal Civil and Administrative Penalties in Australia* (Report 95, 2002) para 4.33.

[14] AI Ogus and C Abbot, 'Sanctions for Pollution: Do We Have the Right Regime?' (2002) 14 *JEL* 283.

[15] DEFRA, *Review of Enforcement in Environmental Regulation, Report of Conclusions* (October 2006), para 5.12.

[16] As discussed in ch 3, a flexible enforcement approach of any type depends on their being an effective and credible deterrent threat.

[17] The enforcement strategies of selected environmental regulators are discussed further in ch 5.

[18] B Wootton, *Crime and Criminal Law*, 2nd edn (London, Sweet and Maxwell, 1981) 47.

But does this particular argument justify the use of strict liability in pollution control offences? The primary aim of the majority of criminal laws is to provide for the conviction of those who cause harm to other citizens or society, although some traditional crimes are also aimed at punishing acts that create the risk of harm. Just as with more established forms of criminal conduct, failure to comply with pollution control legislation can have serious implications for public welfare and safety. In fact, it is difficult to distinguish environmental harms from other more traditional types of criminal damage. Some would argue, for example, that environmental harm does not respect geographical boundaries. But crimes such as internet pornography and drug trafficking also fail to recognise state borders. It could also be argued that, in contrast with other types of criminal harm, remediation of the damage caused by the pollution is difficult, costly and maybe even impossible. Whilst the destruction of an ecologically priceless wetland or the extinction of an endangered species are indeed irreversible, not all environmental damage is of such magnitude and in any event, irreversibility is not a defining aspect of environmental harm. However, what does distinguish environmental pollution from traditional criminal conduct (and ultimately triggers a greater emphasis on risk reduction) is the temporal dimension of environmental harm, and the long-term significance of environmental harm on future generations. Coupled with scientific uncertainty as to the impact of anthropogenic activities on our environment, the stakes are much greater than those presented by more traditional crimes. As aptly put by Cane:

> [I]t is much more important for the future of the human race that we protect the environment than that we reduce the number of road accidents.[19]

We can therefore make a strong case for strict liability on this basis alone.

Similar views have been expressed by the courts. For example, in *Environment Agency v Milford Haven Port Authority (The Sea Empress)*[20], the Court of Appeal (UK) was required to consider whether the defendant Port Authority had caused or knowingly permitted trade or sewage effluent to be discharged into the sea. Lord Bingham CJ, in considering the nature of environmental liability, commented that:

> Parliament creates an offence of strict liability because it regards the doing or not doing of a particular thing as itself so undesirable as to merit the imposition of criminal punishment on anyone who does or does not do that thing irrespective of that party's knowledge, state of mind, belief or intention.[21]

Despite the strong law and economics case in favour of strict liability, and the fact that it is clearly in the public interest to protect the environment, making environmental crime (or indeed any crime) punishable on the basis of strict liability is a contentious issue. Many of the arguments put forward by those who

[19] P Cane, 'Are Environmental Harms Special?' (2001) 13 *JEL* 3, 7.
[20] [2000] 2 Cr App R (S) 423.
[21] *Ibid* 432.

object to its use in criminal law are in fact very persuasive and force us to consider, in more depth, the case for strict liability in the environmental context. One of the most commonly cited criticisms of strict liability is that it is dissonant with the basic premise of criminal liability; that it is a fundamental principle of the common law that no act is punishable in the absence of a degree of criminal intent. Strict liability therefore represents a severe departure from such traditional principles, a departure which, many would argue, is unjustifiable on the basis that the conviction, or indeed the imprisonment of a person who cannot be held to be morally culpable for his acts or omissions, is both a pointless and harmful procedure. As summarised by Smith and Pearson:

> [Strict liability] is pointless, because if there has been no fault, it cannot even act as a meaningful encouragement towards greater care, and it is harmful because it brings blameless people, and the law, into disrepute.[22]

That strict liability is an unwarranted departure from this basic premise of criminal law is probably strongest in its application to individual law-breakers. But what are its merits when considering corporate criminal liability? It is a common assertion that the criminal law was and is fundamentally premised on the notion of individual responsibility and is not therefore well equipped to deal with social organisations such as the company. As a result, determining when a company should be criminally liable for the acts of its employees or officers is a complex one.[23] However, it does not necessarily follow that it would be unfair to make all corporate violators responsible for their behaviour, in the absence of fault. In the case of companies, Ashworth argues that they:

> [W]ield such power (in terms of economic resources and influence), that there is no social unfairness in holding them to higher standards than individuals when it comes to criminal liability—so long as fair warning is given, since companies are run by individuals.[24]

In fact, the development of strict liability offences in the mid-19th century can be partly attributed to the need to deal with corporate deviance. In examining Carson's work on the emergence and enforcement of the Factory Acts[25], Norrie provides an interesting explanation as to the sociological reasons why strict liability was applied to corporations at this time. As was common under the Factory Acts, mens rea was removed from the definition of the offence and inserted into the prosecution practices of the inspectors. This dealt with a number of problems with regard to enforcement. First, those who fell within the remit of the Factory Acts were:

> [M]en not on the periphery of moral life, such as displaced or poverty-stricken workers but men who were at the centre of the emerging political and social order.[26]

[22] M Smith and A Pearson, 'The Value of Strict Liability' (1969) *Crim Law Review* 5.

[23] As discussed in s IV below.

[24] Ashworth above n 12, 167.

[25] See, eg, W Carson, 'Some Sociological Aspects of Strict Liability and the Enforcement of Factory Legislation' (1970) 33 *MLR* 396.

[26] A Norrie, *Crime, Reason and History: A Critical Introduction to Criminal Law*, 2nd edn (London, Butterworths, 2001) 85.

These offenders were therefore peers of those sitting on magistrates' benches, creating an ideological conflict whereby magistrates refused to convict their own kind of criminal offences. The use of strict liability crimes removed this conflict as the offences were not perceived as real crimes. And by inserting the mens rea element into the enforcement practices of factory inspectors, the system avoided the possibility of there being too many prosecutions,[27] whilst at the same time ensuring that there were an adequate number of successful prosecutions to meet the goals of compliance and deterrence.

Many environmental offences that result in prosecution are indeed committed by corporate entities and strict liability, in its application to these bodies can and has been strongly defended, particularly where the regulatory authority adopts a compliance approach to enforcement. As Ashworth states:

> Prosecution for corporate crime often results from the enforcement strategy of a regulatory agency that has been monitoring the company's operations for some time, that sees compliance with the law as its ultimate aim, and that regards criminal prosecution as a last resort. Such an approach, combined with strict liability when the law is enforced, may be defended as an effective crime prevention strategy.[28]

But statistics on the composition of UK business, for example, reveal that 99.3 per cent of all businesses are small (between 0 and 49 employees), and 72.8 per cent of businesses have no employees at all.[29] Ashworth's argument is much less persuasive when applied to these 'micro' SMEs who have neither the economic resources nor in-house expertise to ensure compliance with the maze of complex (and costly) environmental controls. The majority of Environment Agency prosecutions are brought against such persons: the use of strict liability is less easily defended in its application to these micro businesses. To counter these concerns, the potential harshness of strict liability is mitigated by the enforcement practices of environmental regulators and the fact that offenders will not normally be pursued through the criminal courts in the absence of some form of mens rea, whether it is intent, recklessness, negligence or knowledge.[30] However, there is an obvious danger in giving broad discretion to quasi-regulatory authorities such as the Environment Agency in England and Wales. When authorities have the power to put in motion the machinery of the criminal law, their decision-making must be transparent and they must accountable for their actions. This is even more important when we remember the fundamental role that environmental regulators play in determining the scope of liability through the setting of environmental standards. It is therefore of paramount importance that regulators are

[27] This is referred to by Norrie as the problem of 'endemic criminality'. *Ibid.*

[28] A Ashworth, 'Towards a Theory of Criminal Legislation' (1989) *Criminal Law Forum* 41, 52.

[29] See further Small Business Service, *SME Statistics 2005* (available at <http://stats.berr.gov.uk/smes/200612/index.asp> accessed 5 January 2009).

[30] This is reflected in the enforcement and prosecution policies of selected environmental regulators in Australia, Canada and England/Wales. See further ch 5 s II B and C, s III B and C and s IV.

transparent in their prosecutorial decision-making and are effectively held to account.[31]

A traditional argument used to defend strict liability was that the punishments for regulatory offences were light, often trivial. This, however, is no longer the case, with most environmental (and other) offences punishable by a substantial fine and/or imprisonment. Where such grave penalties (with such grave consequences) can be imposed, some would argue that the imposition of liability in the absence of fault breaches notions of fairness and legitimacy. However, this again ignores the fact that as the courts have regard to the culpability of the offender in passing sentence this may lessen the potentially harsh consequences of imposing severe criminal sanctions in the absence of fault.[32] In fact, the English courts have used the seriousness of the offences as an argument in favour of strict liability.[33] This will inevitably result in no-fault liability for some imprisonable crimes.

But if one were to embrace the notion that combining strict liability with potentially severe criminal penalties is unfair, the argument would apply most strongly to individuals who, in contrast with companies (a) can be imprisoned and (b) are less able to absorb any financial penalty imposed. But this is not an intractable problem. As illustrated in Section III, legislators in Australia and Canada have recognised the potential harshness of penalties that are applied to individual offenders found guilty of strict liability offences. Not only can liability be avoided by proving one or more available defence, but also corporate defendants can be subjected to higher penalties than individuals. Although there are no such similar provisions in UK environmental regulations, the Sentencing Advisory Panel is very clear in stating that:

> [T]o cross the custody threshold, a case would need to combine serious damage, or the risk of serious damage, with a very high degree of culpability on the part of the offender.[34]

Despite concerns about the application of strict liability in pollution control (and other) regulatory offences, the relative ease of prosecuting strict liability offences is warmly welcomed by environmental regulators and public prosecutors. However, from a practical perspective, it is often argued that although strict liability should ensure that regulation has a high impact on operations, it actually reduces the likelihood of strong deterrence by trivialising offences. In other words, strict liability goes some way to explaining the disproportionately low level of fines that are often imposed subsequent to conviction.[35] Low fines weaken the extent to

[31] There has been some criticism of Environment Agency transparency and accountability. See, eg, D Bell and T Gray, 'The Ambiguous Role of the Environment Agency in England and Wales' (2002) 11 *Environmental Politics* 76.

[32] For details of sentencing criteria see ch 7 s II C.

[33] See s C below.

[34] Sentencing Advisory Panel's Advice to the Court of Appeal, *Environmental Offences* (1999) para 29 available at <http://www.sentencing-guidelines.gov.uk/advice/index.html#enviro> accessed 5 January 2009.

[35] See further ch 7 s III D. The DEFRA review concludes, however, that the evidence for a 'trivialisation' effect is not strong, and may in fact be diminishing as society increasingly expects the courts to get tougher on polluters. See DEFRA Report above n 15 paras 4.13–4.16.

which the criminal law can induce desirable behaviour, namely compliance with pollution control regulation. Another possible explanation for the low fines imposed on regulatory offenders is that the courts do not view regulatory offences as real crimes. For many years academic circles have debated the extent to which regulatory crimes (in part identifiable by strict liability provisions) can and should be viewed as 'real' crimes. In the next section, I consider this debate, focusing on whether or not the distinction between regulatory and real crime can, in fact, be justified.

C The Real v Regulatory Crime Debate

As a society, we associate the criminal law with what have been termed 'real' or 'traditional' crimes: causing death and wounding, theft, deception and possession of an offensive weapon are just a few examples of behaviour which has been criminalised, in many cases, for hundreds of years. Although many 'real' crimes have their origins in the common law, the bulk are found scattered in numerous statutes with Parliament, over the years, creating a range of new crimes such as stalking, intimidating a witness and most recently, inciting terrorism.[36] Since the mid 20th century, there has also been an increase in the number of legislatively enacted 'regulatory' or 'public welfare' offences.[37] These offences relate to matters such as environmental pollution and health and safety and, unlike 'real' crimes, are enforced by a regulatory authority with the specific remit of promoting compliance with the legislative controls. A question which has generated much interest in academic circles is the extent to which these regulatory offences can be reconciled with the popular notion of crime. Although it is neither necessary nor desirable to rehearse all these arguments now, I would like to make reference to two interesting reasons which have been put forward for justifying the distinction that is made between regulatory and 'real' crime.

First, it has been argued that 'real' crime is concerned with serious anti-social behaviour that offends a sense of common morality: '[T]he stigmatisation and punishment [that are] consequent upon a finding of guilt for a criminal offence require a moral justification.'[38] Regulatory offences, in contrast, are not. In

[36] The distinction between common law and statutory crimes is reflected in the use of the terms *mala in se*, conduct which is inherently criminal and *mala prohibita* ('*quasi*' crimes), conduct which is criminal only because the law has decreed it so. Offences such as murder, rape and theft fall within the *mala in se* category whereas environmental regulation, health and safety laws and food safety legislation eg are 'quasi' crimes. See further C Wells, *Corporations and Criminal Responsibility*, 2nd edn (Oxford, OUP, 2001) 6–7 for further analysis.

[37] Describing regulatory offences as 'public welfare' offences is, in itself, problematic: their characterisation as such 'obscures rather than illuminates' (Wells *Ibid* 6). As argued by Wells, the use of the term reinforces the sense that 'they [regulatory offences] are not only subject to a different administrative and enforcement framework but that they are of a different order from ordinary crime.' (*Ibid*) But are more traditional crimes such as murder, rape and assault not also concerned in some way with public safety and welfare? Do we not prosecute crimes such as murder and rape through the state system of enforcement on the basis that it is in the public interest?

[38] M Bagaric, 'The 'Civil-isation' of the Criminal Law' (2001) *Criminal Law Journal* 184, 185.

response, Ashworth notes that the perception that the criminal law is only concerned with serious anti-social conduct 'must be abandoned as one considers the broader canvas of criminal liability.'[39] Criminal acts such as parking in a 'no parking' zone, graffiti and cycling without lights have, he says, diluted any element of stigma to almost vanishing point. And what about regulatory environmental offences such as polluting a river with toxic substances or fly-tipping asbestos in a protected area? Surely, on the basis of moral repugnance, environmental crimes are 'real' crimes. Fuelled by the growth in the green movement and the recognition that anthropogenic activities can have a hugely negative affect on the quality of our surrounding environment, there is a new widespread public belief that we need to react strongly to environmental abuse. In any event, classifying 'real' crimes as those based on a common sense of morality also raises difficult questions about the extent to which morality can actually be defined and measured.[40] What is morally wrong is dependent, to a great extent on the historical and cultural context in which morality judgements are made.

Second, unlike 'real' crime which requires that the prosecution prove not only the commission of an unlawful act (actus reus) but also some degree of moral responsibility (mens rea) such as recklessness or negligence, regulatory offences may, as discussed in the previous section, often dispense with a mens rea requirement and rest on the notion of strict liability. This has, in itself, further added to the perception that regulatory offences are not crimes in the real sense. But of course, the imposition of strict liability may in fact demonstrate the opposite; that the seriousness of the actual and potential harm resulting from non-compliance with the regulatory controls necessitates a strict liability regime. This argument has, in recent years, been recognised by some English judges hearing environmental cases. For example, in *R v Anglian Water Services*[41], a case involving the discharge of sewage effluent into a watercourse, the defendant water company attempted to distinguish between offences that do not require mens rea and acts of a criminal nature. In response, the Court of Appeal submitted that:

> We would not categorise breaches of section 85(3) [of the Water Resources Act 1991] of the nature that occurred in this case as being of a non-criminal character, albeit the offence is one of strict liability. . . . [P]arliament has imposed on people like the appellant a heavy burden to do everything possible to ensure that they do not cause pollution.[42]

D Conclusion

In this section, I have examined two important issues that arise as a result of criminalising polluting behaviour. Rather than trying to explain how regulatory offences do not necessarily go against traditional criminal law principles, it is

[39] Above n 12, 1.
[40] For a general discussion, see Ashworth *Ibid* 40–44.
[41] [2003] EWCA Crim 2243.
[42] *Ibid* [13]–[14] (Scott-Baker LJ).

better to see the practice of criminalising regulatory environmental offences as introducing a completely different perspective into the criminal legal system. Yes, it is inevitable that the integration of the criminal law into regulatory frameworks established to tackle environmental degradation will prompt the re-examination and reform of a range of existing criminal law doctrines. But whether or not academics enter into such a discourse, using the criminal law to underpin environmental offences is indeed the only apposite response to growing environmental degradation:

> While traditional criminality poses a menace to social stability, environmental crimes do not so much threaten social stability as pose a threat to the survival of human beings to themselves.[43]

The arguments both for and against strict liability are persuasive. On the one hand, the economic and public interest case for strict liability is strong. On the other, the use of strict liability in criminal law is contentious, particularly when applied to individuals. In recognising this, legislators in jurisdictions such as Canada and Australia have, as discussed in Section III below, tiered environmental offences according to the seriousness of the offence and the degree of mens rea required for conviction. So, for example, polluting a watercourse can be prosecuted not only as a strict liability offence but also as an offence requiring proof of intention or negligence. The more serious the offence, the higher the fine that can be imposed on the offender. Depending on the circumstances of each case (and the available evidence) public prosecutors therefore have discretion in determining the most appropriate level at which to prosecute an offender. This, I would argue, is an appropriate way of mitigating the potential harshness of strict liability whilst at the same time harnessing the procedural advantages of no-fault offences. Section III considers rules of liability in Australian Canadian and English pollution control regulation.

III Rules of Liability in Australia, Canada and England and Wales

A Rules of Liability in Australian Pollution Control Regulation

The nature of criminal environmental liability in Australia has been hotly debated for some years, with one judge in the NSW Court of Criminal Appeal acknowledging that the mens rea component of the criminal law has been 'the subject of a bewildering diversity of judicial opinion over the past century.'[44] The starting

[43] AV Noguera, 'Some Reflections on the Use of the Criminal Law for the Protection of the Environment' [2003] 3 *Environmental Liability* 107, 108.

[44] Per Hunt CJ in *EPA v N* (1992) 26 NSWLR 352. Cited in G Bates, *Environmental Law in Australia*, 5th edn (Australia, Butterworths LexisNexis, 2002) 206.

point is to consider the case of *He Kaw Teh v R*, in which the High Court of Australia confirmed the common law presumption that mens rea is an essential element of all criminal offences unless that presumption is displaced either by the words of the statute creating the offence, or the subject matter with which the statute is concerned.[45] The court concluded that there are three approaches to the mens rea requirement in statutory offences: offences where conviction is dependent on proof of intention, strict liability offences where there is no need to prove intent but where defences are available, most notably the common law defence of 'honest and reasonable mistake of fact', and absolute liability offences where no defence can be pleaded.[46] Pollution control regulation in Australia utilises all three approaches.

i *Mens Rea Offences*

Let us begin then with mens rea or fault/based offences. A number of statutes create offences which are, for example, committed wilfully, negligently, intentionally or recklessly.[47] Another common ingredient of fault-based liability is to 'permit' or 'knowingly permit' an offence.[48] This has been interpreted as meaning to 'intentionally allow'[49] and implies a level of control over the unlawful act.[50] Taking reasonable precautions and exercising due diligence to prevent the commission of an offence is often a statutory defence to fault-based pollution control offences.[51] Successfully invoking this 'due diligence' defence will demonstrate lack of negligence or fault in the commission of the offence. However, its application in the context of environmental crime is somewhat uncertain. Take, for example, section 118 of the Protection of the Environment Operations Act (PEOA) 1997 (NSW), which provides a defence to the wilful or negligent disposal or escape of polluting substances of waste. The section states that it is a defence if the person establishes:

> (a) that the commission of the offence was due to causes over which the person had no control; and
> (b) that the person took reasonable precautions and exercised due diligence to prevent the commission of the offence.

Bearing in mind that the due diligence defence assumes that the prosecuting authority has proven wilful or negligent behaviour, it is difficult to comprehend

[45] (1985) 157 CLR 523.

[46] For a summary see Australian Law Reform Commission Report 95 above n 13 at ch 4.

[47] For example, Protection of the Environment Operations Act 1997 (PEOA 1997 (NSW)) ss 115–17 create offences of wilfully or negligently disposing of waste. Also see Environment Protection Act 1970 (EPA 1970) (Victoria) s 59E.

[48] Knowledge of what is being permitted is an essential element of the offence, so it matters not whether the legislation states 'permit' or 'knowingly permit.'

[49] *Environment Protection Authority v Multiplex Constructions P/L* (2000) 112 LGERA 1, 54 (Lloyd J).

[50] See further *Alphacell v Woodward* [1972] AC 824, discussed in s C below.

[51] See eg s 118 PEOA 1997 (NSW).

how the offence could be caused by something over which the defendant had no control. The defence is therefore difficult to make out.[52]

ii Strict Liability Offences

There are two primary models of strict liability offences. First, if a statutory provision does not refer to mens rea, then the courts will generally presume that the offence is one of strict liability, on the basis that environmental crimes are societal crimes which, by nature, warrant the imposition of strict liability. So for example, section 120(1) of the PEOA 1997 (NSW) states that 'a person must not pollute any waters'. There is no explicit mention of a mens rea component and so the offence is deemed to be one of strict liability. The use of the word 'cause' in pollution legislation[53] also indicates that liability is strict, with the Australian courts adopting an approach similar to that of the House of Lords in the *Empress Car Company* case.[54] Second, legislation may explicitly state that proof of mens rea is not required. So for example section 196 of the federal Environment Protection and Biodiversity Conservation Act 1999 states that strict liability applies to the offence of killing or injuring a member of a listed threatened species.

As noted earlier in this section, the common law defence of honest and reasonable mistake of fact applies to strict liability offences:

> [I]f it is held that guilty knowledge is not an ingredient of an offence, it does not follow that the offence is an absolute one. A middle course, between imposing absolute liability and requiring proof of guilty knowledge or intention, is to hold that an accused will not be guilty if he acted under an honest and reasonable mistake as to the existence of facts, which, if true, would have made his act innocent.[55]

To invoke the common law defence of honest and reasonable mistake of fact, the defendant has to prove that the offence was committed under a mistaken but reasonable belief about facts that, had they existed, would have meant that the conduct was lawful. In other words, the defence involves absence of a guilty mind.[56] It is clear that in order to raise the defence, the mistaken belief must be something more than inadvertence or a mere absence of knowledge.[57]

[52] Demonstrating that the defendant has done her best in the circumstances to prevent the pollution would not normally be enough. (See eg *State Pollution Control Commission v Kelly* (1991) 5 ACSR 607) General precautions will not suffice and the courts will usually look for a comprehensive environmental management system that has been communicated to the workforce. See further Bates above n 44, 215–16.

[53] See eg PEOA 1997 (NSW) s 120(2), which states that 'a person must not cause any waters to be polluted'.

[54] [1998] 1 All ER 475. See further s C below.

[55] *He Kaw Teh* above n 45, 533 (Gibbs CJ).

[56] Where a defendant can provide sufficient evidence to indicate that the defence may apply, then the burden shifts and the prosecution must disprove the defence beyond reasonable doubt.

[57] *Environment Protection Authority v Tyco Water Pty Ltd.* [2005] NSWLEC 453. Some doubt has been cast over whether the common law defence can be pleaded where a Tier 2 offence has been committed under the EPA 1970 (Victoria). Past decisions have indicated that the defence is unavailable,

Environmental legislation may obviously also include its own statutory defences, common examples being that the action constituting the offence was authorised under a valid environmental licence,[58] that the offence occurred as a result of an emergency,[59] that the commission of the offence was one over which the accused had no control, or that the offender acted with 'due diligence'.[60] In States with a Criminal Code, it is a matter of interpretation as to whether the defences in the Code apply to the environmental legislation in question.[61]

iii Absolute Liability Offences

An absolute liability offence is defined as such by virtue of the fact that the defence of honest and reasonable mistake is not available. Examples are, however, rare, and where an absolute liability offence is available it will carry very small penalties. As noted by Bates, absolute liability most commonly occurs in the case of marine pollution from ships and other vessels where legislation clearly sets out a range of specific statutory defences.[62]

iv Tiering Offences

Perhaps in recognition of the potential harshness of strict liability, especially where the penalties can be large, many states 'tier' offences according to the nature of liability and/or the seriousness of harm. In Victoria, maximum penalties for crimes requiring mens rea are generally higher than for strict or absolute liability crimes. For instance, the Environment Protection Act 1970 (the EPA 1970) divides environmental offences into three broad categories, according to the degree of culpability and the likely seriousness of environmental harm. The most serious first tier offences are defined in section 59E of the Act as those of 'aggravated pollution' where an offence is dependent on proof of intention, recklessness or negligence. So, a person who intentionally, recklessly or negligently pollutes the environment, resulting in serious environmental damage, attracts the most severe penalties, including imprisonment.[63] The EPA 1970 also includes a range of second tier,

thereby suggesting that the offences are absolute as opposed to strict liability. (See *Allen v United Carpet Mills* [1987] VR 323, followed in *Wilson v Gahan* [1999] VSC 72.) But since the dramatic rise in penalties applicable to second tier offences in Victoria, the rationale behind these decisions may no longer be applicable. See further W Gumley, 'Environmental Crimes: Offences and Penalties in Victoria' published proceedings of the Corporate Crime Workshop, February 2002, 111.

[58] Eg s 122 PEOA 1997 (NSW).

[59] For example, s 30B EPA 1970 (Victoria).

[60] S 118 PEOA 1997 (NSW).

[61] For further details, see Bates above n 44 ch 9 s 9.9–9.10.

[62] These offences reflect international marine oil pollution agreements.

[63] Whilst the offence clearly requires proof of *mens rea*, there is some doubt as to whether the state of mind extends to other elements, eg, the serious environmental damage resulting from the offence. See further, G Bates and Z Lipman, *Corporate Liability for Pollution* (Sydney, LBC Information Services, 1998) 140.

moderate offences. For example, actions that pollute water, air and land are prohibited under the Act.[64] These offences do not explicitly require proof of mens rea, be it intention, recklessness or negligence although, in some cases, a higher penalty may be imposed for second tier offences committed with intent.[65] Finally, third tier 'minor' offences, such as failing to report an incident, are generally treated as offences of absolute liability and are usually dealt with by way of infringement notices or a small fine.

A similar system is adopted in NSW under the PEOA 1997, although arguably the NSW approach is based more on levels of proof and less on the seriousness of harm. Tier 1 offences such as disposing of waste in a manner likely to harm the environment are the most serious and require proof of mens rea, notably wilfulness or negligence.[66] Tier 2 offences are defined as all other offences under the act or associated regulations and fall into three broad categories: causing and permitting pollution or environmental harm, offences pertaining to administrative breaches (such as non-compliance with an environmental protection notice) and offences relating to enforcement. As noted earlier, due to the fact that these offences make no explicit reference to intention, recklessness or negligence, they are generally deemed to be provisions of strict liability. Tier 3 offences are classed as Tier 2 offences that can be dealt with, under the Act, by way of a penalty notice (an on-the-spot fine).[67]

B Rules of Liability in Canadian Pollution Control Regulation

Environmental offences in Canada are generally described as regulatory or public welfare offences as opposed to 'true crimes'. Mr Justice Cory in *R v Wholesale Travel Group Inc* recognised this distinction and the resulting nature of liability:

> The objective of the regulatory legislation is to protect the public or broad segments of the public . . . from the potentially adverse effects of otherwise lawful activity. Regulatory legislation often involves a shift of emphasis from the protection of individual interests and the deterrence and punishment of acts involving moral fault to the protection of public and societal interests . . . Since regulatory offences are directed primarily not to conduct itself but to the consequences of conduct, conviction of a regulatory offence may be thought to import a significantly lesser degree of culpability than conviction of a true crime. The concept of fault in regulatory offences is based upon a reasonable care standard and, as such, does not imply moral blameworthiness in the same manner as criminal fault. Conviction for breach of a regulatory offence suggests nothing more than the defendant has failed to meet a prescribed standard of care.[68]

[64] See ss 39, 41 and 45 respectively.
[65] See s 67AA.
[66] See PEOA 1997 (NSW) ch 5 Pt 5.2. For details see R Lyster, Z Lipman, N Franklin, G Wiffen and L Pearson, *Environmental and Planning Law in New South Wales* (Sydney, The Federation Press, 2007) 440–67. Also see ch 7 Table VI.
[67] S 114(3) PEOA 1997 (NSW).
[68] [1991] 3 SCR 154, 219.

As in both Australia and England and Wales, regulatory offences in Canada can be divided into three categories: mens rea, strict liability and absolute liability. A few environmental offences can be classed as mens rea offences under which the Crown must prove beyond reasonable doubt that the defendant committed the offence and had the requisite guilty mind in the form of intention, knowledge or recklessness.[69] At the other end of the spectrum, where an offence is based on absolute liability, the accused cannot exonerate himself by showing that he was free of fault. There are very few absolute liability offences in Canadian environmental law due to the fact that the Canadian Charter of Rights and Freedoms is infringed where an absolute liability offence is punishable by imprisonment: most federal and provincial environmental statutes allow offenders to be jailed. However, recent amendments to the Environmental Protection Act (EPA) 1990 (Ontario) introduce absolute liability for certain offences. For example, where an offence is committed under the Act for which an environmental (administrative penalty) is payable, liability is absolute. It is no defence that the defendant took all reasonable steps to prevent the contravention or had an honest and reasonable belief in a mistaken set of facts that if true would have rendered the contravention innocent.[70]

As identified by Mr Justice Cory (above), the vast majority of environmental offences are strict liability offences. For example, under section 40 of the EPA 1990 (Ontario), 'no person shall deposit or cause, permit or arrange for the deposit' of waste on land except under and in accordance with the terms and conditions of a certificate. More generally, under section 186 every person who contravenes the Act is guilty of an offence. Higher fines are payable where an offence is committed by a corporation. Whether statutorily provided for or not, the presence of a 'due diligence' defence distinguishes absolute and strict liability offences.[71] In other words, the accused can avoid strict liability offences by proving that he took all reasonable care/exercised due diligence. Taking all reasonable care involves consideration of what a reasonable man would have done in the circumstances. More specifically:

> The defence will be available if the accused reasonably believed in a mistaken set of facts which, if true, would render the act or omission innocent, or he took all reasonable steps to avoid the particular event.[72]

What constitutes due diligence in each case is a question of judicial interpretation. Judge Ormston of the Ontario Provincial Court in *R v Bata Industries Ltd* sets out a useful checklist of what may be evidence of due diligence.[73] He begins by considering whether the Directors of the company in question established a pollution prevention system. He then considers inter alia the extent to which employees were

[69] For example, CEPA 1999 s 274(1)

[70] S 182.1(6). Contraventions for which an environmental penalty is payable include breach of a discharge limit or breach of a permit or licence.

[71] For a detailed analysis see D Saxe, *Environmental Offences: Corporate Responsibility and Executive Liability* (Ontario, Canada Law Book Inc, 1990) ch 6.

[72] *R v Sault Ste Marie (City)* [1978] 2 SCR 1299, (1978) 40 CCC (2d) 353, 374 (Dickson J).

[73] (1992) 9 OR (3d) 329.

made aware of this system, the existence of monitoring and reporting on environmental issues, the speed with which environmental concerns were addressed, and the response of Directors when the system failed.[74]

Canadian pollution control statutes are not, however, always explicit in the nature of liability imposed so it is often down to judicial interpretation. According to Benedickson, an offence will be classed as requiring mens rea where the language used in the statutory provision indicates that the accused's state of mind is an important factor to be established by the prosecution.[75] For example, words such as wilfully, knowingly or with intent are relatively clear in requiring mens rea. Distinguishing between absolute and strict liability can, however, be more difficult. Where there are no specific statutory due diligence provisions, the classification of an offence as one of strict liability is a matter of interpretation. Dickson J's judgement in the seminal case of *R v Sault Ste Marie* paved the way for subsequent interpretations of environmental law statutes.[76] The case itself involved a breach of section 32(1) of the Ontario Water Resources Act which stated that every municipality or person that discharges or deposits or causes or permits the discharge or deposit of any material into any watercourse is guilty of an offence. Dickson J began his analysis by stating that public welfare offences would prima facie impose strict liability and that absolute liability would apply where it was clear that the legislature intended that guilt would follow proof merely of the proscribed act, taking into account, inter alia, the subject matter of the legislation, the overall regulatory pattern, the nature of the penalty imposed and the precision of the language used.[77] He noted that words such as 'cause' and 'permit' fitted much better into an offence of strict liability and in, the absence of any clear indication that liability was absolute and without any words such as 'knowingly', applying the criteria above, the offence was one of strict liability.[78]

C Rules of Liability in England and Wales

Generally speaking, and unless the statutory provision indicates otherwise, the courts in England and Wales will interpret statutory offences on the basis that an element of mens rea is required. However, most pollution control offences impose strict liability.[79] For example, consider the oft-litigated section 85(1) of the Water Resources Act 1991, under which it is an offence to 'cause or knowingly

[74] *Ibid* [146].
[75] J Benidickson, *Essentials of Canadian Law: Environmental Law*, 2nd edn (Toronto, Irwin Law, 2002) 141.
[76] Above n 72.
[77] *Ibid* [61].
[78] *Ibid* [66].
[79] *Sweet v Parsley* [1970] AC 132. This is in sharp contrast to the criminal environmental law systems of other European member states where liability is mostly dependent on proof of negligence or intent. See further *Criminal Penalties in EU Member States' Environmental Law* co-ordinated by Professor Michael Faure and Professor Günter Heine. Available at <http://ec.europa.eu/environment/docum/pdf/02544_final_report.pdf> accessed 7 January 2009.

permit poisonous, noxious or polluting matter or any solid waste matter to enter any controlled waters' except under and in accordance with a consent. In the case of *Empress Car Co v National Rivers Authority*,[80] the House of Lords was confronted with determining the nature of liability under section 85, more specifically the meaning of the phrase 'to cause'. In finding the company liable, Lord Hoffman considered the nature of the duty imposed by section 85(1):

> Does it [liability] include responsibility for acts of third parties or natural events and, if so, for any such acts or only some of them? This is a question of statutory construction, having regard to the policy of the Act. It is immediately clear that the liability imposed by the subsection is strict: it does not require mens rea in the sense of intention or negligence. Strict liability is imposed in the interests of protecting controlled water from pollution.[81]

So, what does have to be proved in order for a defendant to be successfully convicted of causing pollution under section 85? In answering this question, the court followed the decision in the earlier case of *Alphacell v Woodward*, where the defendant owner of paper making mills was found liable, in the absence of negligence, for allowing effluent to pollute a local river.[82] In the *Alphacell* case, it was held that as 'the whole complex operation which might lead to [pollution] was an operation deliberately conducted by the appellants' they, in the absence of an extraordinary event, caused the pollution.[83] In the *Empress* case, Lord Hoffman reiterated this principle, stating that in order to 'cause' pollution, the defendant must have done something, such as 'maintaining tanks, lagoons or sewage systems full of noxious liquid'.[84] He did, however, reject the notion of absolute liability, stating that a defendant was not liable for the escape of polluting matter, irrespective of how this happened. The causal effects of the defendant's acts would be negated where the escape was due to the 'extraordinary' acts of a third party or 'extraordinary' natural events. The particular form of vandalism that led to the release of the oil in this case was not extraordinary and the defendant company was liable. In other words, '[B]y indulging in the regulated activity, the offender has voluntarily adopted the risks of regulatory infraction . . .'.[85]

In some instances, the term 'cause' is preceded by the term 'knowingly', the use of which could be an explicit way of requiring mens rea as to certain elements of

[80] [1999] 2 AC 22. Briefly, the facts of the case are as follows. The appellant company had installed a large diesel tank in its yard that drained directly into a river. To contain spillages, the tank was surrounded by a bund wall. However, at the time of the incident, an extension pump was running from the tap on the diesel tank to a smaller drum which was located outside the bunded area. One night, an unidentified third person opened the tap, thereby allowing the contents of the tank to make their way into a storm drain and thence into a nearby river. The company was charged with 'causing' pollution of the river.

[81] *Ibid* 32 (Hoffman LJ).

[82] [1972] AC 824.

[83] *Ibid* 834.

[84] Above n 80, 35 (Hoffman LJ).

[85] G Richardson, 'Strict Liability for Regulatory Crime: the Empirical Research' [1987] *Crim LR* 295, 295–6.

the offence. However, in the first decision on the meaning of the word 'knowingly' as found in section 33 of the Environmental Protection Act 1990, the High Court concluded that the test is strict. In the case of *Shanks and McEwan (Teesside) Ltd v Environment Agency*[86], a site supervisor working for the defendant failed to comply with a waste management licence condition stating that a new waste disposal form should be completed where waste was transferred to an alternative location on site. The company was charged with 'knowingly causing' controlled waste to be deposited in breach of a waste management licence condition. The company argued that it could not have known of the breach of the condition because the only person who could have known, ie, the site supervisor, was not part of the directing mind of the company.[87] The High Court held that it was not necessary to prove that the company knew of the breach—the offence was satisfied by the fact that the company knowingly operated and held out its site for the reception and deposit of controlled waste.

Perhaps in response to the potential harshness of imposing criminal responsibility on those who had no knowledge of the act or could not have avoided the act, the strictness of environmental offences in England and Wales is sometimes mitigated by the availability of statutory defences.[88] This is particularly apparent in the case of offences under the waste management licensing regime. Defences include taking all reasonable precautions and exercising due diligence to avoid the commission of the offence.[89] This generally entails that the correct steps were taken in the circumstances or an adequate control system was set up. It is also often a defence where the acts causing the pollution were done in an emergency to avoid danger to human health.[90]

A recent DEFRA review of environmental enforcement, in recommending a review of the sentencing framework, has proposed several reform options, all of which would require amendments to existing liability provisions. In doing so, the review emphasises the fact that, in order to ensure that criminal courts have the greatest effect, 'the reputational penalty from any moral condemnation should be reinforced as far as possible to create stronger deterrence.'[91] The most radical option is to replace all existing criminal offences with new offences based on serious culpability. Other offences would be dealt with by means of variable administrative penalties. Less radical options include retaining strict liability but only in

[86] [1999] QB 333.

[87] For a discussion of the identification or direct liability doctrine, see s IV.

[88] There are however, some exceptions. The defence of due diligence is not available for an offence committed under Water Resources Act 1991 s 85(1) and Pollution Prevention and Control Regulations 2000 reg 32.

[89] See, eg, Environmental Protection Act 1990 s 33(7)(a).

[90] See, eg, Environmental Protection Act 1990 s 33(7)(c). This is qualified by two requirements: all reasonably practicable steps must have been taken to minimise pollution of the environment and harm to human health; and the Environment Agency must be informed of the acts as soon as is reasonably practicable. The word 'emergency' is to be viewed objectively, without any reference to how the licence holder perceives a given set of facts (see *Waste Incineration Services Ltd. v Dudley MBC* [1993] Env LR 29. This case involved an analogous defence under Control of Pollution Act 1974 s 3(4)(d).

[91] Above n 15, para 6.3.34.

cases of very serious damage, economic gain and harm to communities and introducing separate mens rea offences, or amending the existing prosecution policy and stipulating that only cases where mens rea could be proven be brought to the criminal courts. Unfortunately, legislative changes have not yet been introduced.

IV Strengthening Deterrence? Corporate and Director Liability

A Introduction

In this section, I consider the role that corporate and director liability plays in inducing companies and senior managers to comply with environmental regulation. How and when can a company and its directors be prosecuted for committing an environmental offence, and will this strengthen the deterrent impact of enforcement? The purpose of this section is not to engage in a lengthy debate about whether or not corporate criminal liability is desirable, suffice to say that, in the environmental context, the arguments in favour are strong.[92] First and foremost, corporations are regarded as serious sources of potential harm. For example, it is estimated that in the UK, small and medium-sized enterprises are responsible for 60 per cent of industry's carbon dioxide emissions[93], 60 per cent of commercial waste and 80 per cent of all pollution incidents.[94] Coupled with the growing pressure on companies to increase profitability, corporate criminal liability is needed to internalise the costs of corporate endeavours on society.

From a deterrence perspective, will a sanction imposed on a company induce behavioural changes in the individuals of that company? The conventional answer is that, as the primary goal of most companies is to a make a profit, and the success of management is measured by reference to that profit, then managers will be motivated to reduce costs, including those incurred as a consequence of socially undesirable activities (such as pollution) undertaken by employees. As will be illustrated in Chapter 7, targeting a sanction at a firm may indeed be more effective in deterrence terms than prosecuting an individual employee of that firm, particularly where the employee's wealth is insufficient to pay a fine. A firm may also have a variety of internal sanctions that are not available to regulators or the courts such as reduced wages or lower prospects of promotion. Furthermore, there is a presumption that a sanction directed at the firm will impact on specific groups, mainly shareholders who are in a strong position to influence corporate policy and management. Of course, in some cases the cost of a corporate sanction may

[92] For a detailed analysis see Wells above n 36 ch 2.
[93] Marshall Report, *Economic Instruments and the Business Use of Energy* (London, HMSO, 1998).
[94] Environment Agency, *SME-nvironment 2005*, 1.

be passed on to other less influential groups such as employees or consumers but the introduction of 'equity fines', for example, could counter this risk.[95] All in all, despite limitations, empowering regulatory agencies and public prosecutors to bring actions against corporate entities strengthens the deterrent impact of enforcement action.

Assuming corporate liability does induce compliance by individual actors, there are two key differences in how corporate as opposed to individual sanctions affect the deterrence function of the criminal law. First:

> It is easier to target the sanction on [utility/benefit] because the problematic non-financial dimension to the utility derived from the activity does not typically exist in relation to firms.[96]

Corporate accounting requirements should also mean that information on profits is readily available. The second, and most obvious difference, is that some non-financial sanctions, most notably imprisonment, are not available when sentencing corporate entities. Although, as stated earlier, the wealth of companies is likely to be higher than that of individuals, there is still a 'deterrence trap' in fixing penalties high enough to deter. This is further complicated by the fact that the courts will normally try to avoid insolvency. Designing a system of corporate criminal sanctions which can deter and promote good practice is therefore fraught with difficulties.[97]

Although there is widespread support for criminalising corporate polluting behaviour, determining how and when a company will be liable for offences committed by its employees or officers is a complex one, particularly as the framework of criminal liability has historically centred on individual as opposed to corporate defendants. In recognising that corporations have corporate personality (just as natural personality attaches to individuals), judicial and legislative bodies have developed various ways of imposing criminal liability on companies. For example, some statutory provisions may clearly be directed at companies no less than individual employees. Legislation may, for example, impose a duty on employers to exercise reasonable care in operating a process so as minimise the risk of pollution. Alternatively, the doctrine of vicarious ('agency') liability will make a company liable for offences of strict liability committed by employees in the course of employment. It is a matter of statutory interpretation as to whether or not a particular legislative provision attracts vicarious liability and it remains a popular basis of corporate criminal liability, despite the fact that it has been criticised for:

> [I]ncluding too little—in demanding that liability flow through an individual, however great the fault of the corporation—and for including too much—in blaming the

[95] I discuss this in detail in ch 7 s III E.

[96] AI Ogus, *Costs and Cautionary Tales: Economic Insights for the Law* (Oxford, Hart Publishing, 2006) 133

[97] There is some dispute over whether these goals can be served more appropriately by corporate civil liability. For further discussion see VS Kanna, 'Corporate Criminal Liability: What Purpose Does it Serve?' (1996) 109 *Harvard Law Review* 1477.

corporation whenever the individual employee is at fault, even in the absence of corporate fault.[98]

A more conceptual challenge is posed for mens rea offences. Who in a corporation must have mens rea in order for that corporation to have mens rea? To deal with such offences, the courts have developed what has become known as the 'identification' or 'direct' liability doctrine, developed in the case of *Tesco Supermarkets v Nattrass*.[99] Under the *Tesco* principle, a company will be directly liable where the offence is committed by somebody who is part of the 'directing mind and will' of the company, maybe the controlling officers such as the Board of Directors. This test is, however, limited in its ability to achieve corporate convictions. First, only the wrongdoing of a small number of people within a company will be attributed to the company. And second, it is much easier to convict small as opposed to large companies, on the basis that the bigger the company, the more removed the directors will be from the working practices of the corporation. In response to both the problems associated with imposing vicarious liability for criminal offences, and the limitations of the identification doctrine, alternative approaches have been utilised in many jurisdictions.

Regulators also have a second weapon to fight with in the battle to ensure corporate compliance. By 'piercing the corporate veil', individual directorial liability can arguably respond to doubts over the fairness of corporate liability in some contexts and is also possibly better able to influence the behaviour of corporate managers than corporate responsibility. It can also respond to the limitations of corporate sanctions, especially fines.[100] It is therefore not surprising to learn that individual directorial liability is a common feature of pollution regulation in Australia, Canada, and England and Wales and has received much support from law-makers and regulators. It would seem, however, that regulators rarely use their powers to prosecute corporate executives. This, coupled with the closely related fact that attaching liability to these individuals is difficult, means that the potential deterrent impact of directorial liability is relatively weak.

B Corporate and Director Liability in Australia

At the Commonwealth level, the Criminal Code, which applies to corporate entities in the same way as it does to individuals, prescribes the circumstances in which criminal responsibility will attach to corporations. Under section 12.2:

> Where the physical element of an offence is committed by an employee . . . acting within the actual or apparent scope of his of her employment, or within his or her actual or apparent authority, the physical element must also be attributed to the body corporate.

[98] Above n 36, 153.
[99] [1972] AC 153.
[100] For details see ch 7 s III C and D.

For mens rea offences (other than those requiring negligence), an extended version of the Tesco identification doctrine is used (known as the 'corporate culture' model) and a company will be held liable for an offence where it 'expressly, tacitly or impliedly authorised or permitted the commission of the offence'.[101] This authorisation or permission can be established by inter alia proving that the board of directors (or a high managerial agent such as a company officer) carried out or authorised conduct which led to the offence or that the corporate culture directed, encouraged, tolerated or led to non-compliance with the law or that the company failed to create and maintain a corporate culture that required compliance with the law.[102] Due diligence may be a defence in some instances. By using a 'corporate culture' model such as the one described, prosecutors do not have to find an individual who is responsible for the unlawful act: responsibility is placed on the corporate entity as a whole.

In Australian state environmental law, the law has attributed responsibility to corporations in a number of ways.[103] They can be guilty of criminal conduct under the principle of vicarious liability, by which liability will be imposed where the 'intent and purpose of the statute' in question supports the application of the principle.[104] For example, in the case of *Tiger Nominees Pty Ltd v State Pollution Control Commission*, the LEC in NSW concluded that imposing vicarious criminal liability for the acts of employees in the course of their employment was consistent with the purpose and scope of the Clean Waters Act 1970 (NSW).[105] This common law principle has been superseded in many states by legislative provisions which impute vicarious criminal liability. Some provisions extend vicarious liability to both the conduct and state of mind of corporate officers. For example, under section 127 of the Environment Protection Act 1993 (South Australia):

> The conduct and state of mind of an officer, employee or agent of a body corporate acting within the scope of his or her actual, usual or ostensible authority will be imputed to the body corporate.

Sometimes, statutory provisions will utilise a form of the *Tesco* identification doctrine. In NSW, for example, the PEOA 1997 extends the range of persons whose state of mind can be attributed to the corporation to people lower down the corporate ladder, namely officers, employees or agents of the corporation.[106]

At common law, corporate officers are not liable for offences within the corporation. However, environmental protection legislation in all six states and two territories also provides for the personal criminal liability of directors and other senior managers within a company, regardless of whether or not the com-

[101] Criminal Code ss 12.2, 12.3.

[102] Criminal Code s 12.3.

[103] For a more detailed overview, see Law Reform Commission New South Wales, *Sentencing: Corporate Offenders* (Report 102 (2003) ch 2.

[104] Above n 44 (Bates), 218.

[105] (1992) 25 NSWLR 715. The Clean Waters Act 1970 has subsequently been repealed by the PEOA 1997 (NSW).

[106] S 169(4).

pany is prosecuted for the offence (although the prosecution does have to prove that the company committed the offence).[107] Liability arises solely by virtue of the corporate officer's position within the company. For example, under section 169(1) of the PEOA 1997:

> [I]f a corporation contravenes, whether by act or omissions, any provision of this Act or the regulations, each person who is a director of the corporation or who is concerned in the management of the corporation is taken to have contravened the same provision.[108]

Under the original provision in the PEOA 1997 (NSW), three defences were available which, if proven, would exonerate individual officers from liability: where the person could demonstrate that: (a) the company committed the offence without their knowledge (actual, imputed or constructive); or (b) they were not in a position to influence corporate conduct; or (c) they used all due diligence to prevent the offence.[109] However, the availability of the defences, particularly the no knowledge defence, was called into question on the basis that it was out of step with modern notions of corporate responsibility 'which expect managers and directors to be aware of the environmental effects of their businesses'.[110] In order to successfully attach liability to a corporate officer, the prosecutor had to have direct and actual evidence of a director's knowledge of the commission of the offence. The chances were then that individual officers would be exonerated from liability.[111] Howard goes so far as to suggest that the 'no knowledge defence has arguably stemmed the flow of charges laid against directors.'[112] In response to these concerns, the no knowledge defence has been repealed.[113] Similarly, the Victorian Environment Protection (Amendment) Act 2006 repealed the no knowledge defence that had previously been available under section 66B(1A) of the EPA 1970.[114] These changes brought the Act into line with directors' duties and due diligence obligations elsewhere.[115]

In pursuing compliance, state environmental regulators clearly attach great importance to the ability to take enforcement action against companies and indi-

[107] For a detailed (although somewhat dated) overview of director liability see T Howard, 'Liability of Directors for Environmental Crime: the Anything-but-Level Playing Field in Australia' (2000) 17 *Environmental and Planning Law Journal* 250, 259–63.

[108] For a detailed analysis of this provision see Lyster at al above n 66, 461–6.

[109] For a more detailed explanation of the s 169, particularly the defences, see Howard above n 107, 259–63.

[110] Department of Environment and Conservation (NSW), *Report on the Review of the Protection of the Environment Operations* Act (October 2003) 10.

[111] The provisions of s 169 make it clear that the defendant has to prove one or more of the three defences. But in order to bring a charge, the prosecutor, as a matter of practicality, must be satisfied that he/she has evidence of knowledge.

[112] Above n 107, 260.

[113] Protection of the Environment Operations Amendment Act 2005.

[114] Three of the defences set out in the EPA (Victoria) are identical to those provided in the NSW legislation. The fourth defence states that it is a defence if that person can prove that the corporation would not have been found guilty of the offence by reason of its being able to establish a defence available to it under the Act. For details see Howard above n 107, 263–4.

[115] See further Howard *Ibid* 265.

vidual directors, senior managers etc. According to the NSW EPA's Prosecution Guidelines for example, the fact that liability is imposed on a wide range of people is 'in keeping with the aims of the environment protection legislation.'[116] However, as it may not always be appropriate to prosecute every person who may be liable, each regulator will take into account various factors in deciding the most appropriate person/s to take action against.[117] For example, in NSW, action will usually be commenced against a corporation where the offence was committed by 'employees, agents or officers of a corporation in the course of their employment.'[118] This is undoubtedly a reference to vicarious liability, despite the PEOA's use of an expanded version of the *Tesco* doctrine. In the case of directors and managers, the EPA (Victoria) will take into account factors such as 'the contribution of company policy to the incident' and any 'fundamental flaws in the company's systems which contributed to the incident.'[119] Across Australia, director prosecutions would appear to be relatively rare with evidence and commentary suggesting that despite being the most litigious of the Australian states NSW has, compared with Victoria, a relatively poor record in prosecuting directors and other individual corporate officers.[120] So despite the seemingly strong deterrent impact of both corporate and directorial liability, the lack of prosecutions weakens the deterrence effect of this particular enforcement tool.

C Corporate and Director Liability in Canada

In Canada, corporations are clearly expected to comply with environmental regulations. Corporate liability for criminal environmental offences is achieved using a number of different models, although there does appear to be some uncertainty as to the actual mechanisms of accountability. Vicarious liability, despite being rejected by the Supreme Court of Canada in the seminal case of *R v Sault Ste Marie*[121], is the dominant approach used in provincial environmental legislation, particularly for strict liability crimes. So, for example, section 192 of the Environment Protection Act (Ontario) states that:

> [A]n act or thing done or omitted to be done by an officer, official, employee or agent of a corporation in the course of his or her employment or in the exercise of his or her

116 Department of Environment and Conservation/EPA, *EPA Prosecution Guidelines* (July 2004) s C para 5.1 available at <http://www.environment.nsw.gov.au/legislation/prosguide/index.htm> accessed 5 January 2009.

117 See generally s C, *Ibid.*

118 *Ibid* para 6.1.

119 EPA Victoria, *Enforcement Policy* (July 2006) available at <http://epanote2.epa.vic.gov.au/EPA/publications.nsf/d85500a0d7f5f07b4a2565d1002268f3/39a0493666adf794ca2571a2002cdb9e/$FILE/384.3.pdf> accessed 5 January 2009.

120 For examples of prosecutions in all states and territories see Howard. See also RJ Baird, 'Liability of Directors and Managers for Corporate Environmental Offences—Recent Prosecutions' (2000) 17 *Environmental and Planning Law Journal* 192.

121 Above n 72.

powers or the performance of his or her duties shall be deemed to be also an act or thing done or omitted to be done by the corporation.[122]

As noted by Swaigen, the purpose of these provisions is 'to discourage employers from shifting responsibility for violations onto their employees as a way of avoiding corporate responsibility.'[123]

Corporations may also be held liable for mens rea offences under the *Tesco* identification doctrine. This doctrine was endorsed in the *Sault Ste Marie* case[124], in which the Supreme Court stated that a corporation could be convicted if it had caused or permitted pollution to occur by failing to exercise due diligence in the sense of establishing, maintaining and monitoring pollution prevention policies and systems. The court used the identification doctrine to impute liability on the corporation. But the application of the 'directing mind and will' principle has proved to be far from straight-forward. Despite various modifications and clarifications, the Canadian courts have grappled with the problems inherent in establishing the relative roles and responsibilities of various persons within the corporation and in short, prosecutors have been unsure whether to prosecute the company, its employees or even both.[125] As one Ontario MP stated in 1986:

> A popular line of defence for large corporations is to argue they have no legal responsibility for the conduct of their employees and agents. This has made it necessary to prosecute employees and contractors who are in a poor position to defend themselves against such allegations. The purposes of prosecution, such as deterrence, are poorly served when the brunt of legal responsibility falls on people with little power within a corporation.[126]

Due to the perceived weaknesses of the vicarious liability and identification models, some laws at federal and provincial level have, since the late 1980s, been amended with a view to finding a more effective and fair way of attributing criminal responsibility to organisations. A good example of a more recent amendment is the Canadian Criminal Code (which is applicable to all federal legislation).[127] The rules contained in section 22 of the amended Code not only represent a codification of existing common law rules but also go further by broadening the range of individuals whose actions and intentions can trigger corporate liability. For

[122] Also see, eg, s 253 Environmental Protection and Enhancement Act 1992 (Alberta) which states that: 'For the purposes of this Act, an act or thing done or omitted to be done by a director, officer, official, employee or agent of a corporation in the course of that person's employment or in the exercise of that person's powers or the performance of that person's duties is deemed also to be an act or thing done or omitted to be done by the corporation.'

[123] J Swaigen, 'Ontario's Environment Enforcement Statute Law Amendment Act' (1988) 2 *Canadian Environmental Law Association* (N.S.) 14.

[124] Above n 73.

[125] For a more detailed overview, see M Bowden and T Quigley, 'Pinstripes or Prison Stripes? The Liability of Corporations and Directors for Environmental Offences' (1995) 5 *Journal of Environmental Law and Practice* 209, 226–8.

[126] Hansard Official Report of Debates, Legislative Assembly of Ontario, 2nd Session, 33rd Parliament, Tuesday December 16 1986, 4178.

[127] An Act to Amend the Criminal Code (Criminal Liability of Organizations) SC 2003, c 21.

example, in the case of criminal offences which require the prosecution to prove negligence, an organisation will be liable if one of the company's representatives is party to the offence and the senior officer responsible for relevant aspects of that organisation's activities departs markedly from 'the standard of care that, in the circumstances, could reasonably be expected to prevent a representative of the organisation from being a party to the offence'.[128] The term 'representative' is construed widely to include 'a director, partner, employee, member, agent or contractor of the organization.' Similarly, a 'senior officer' is:

> A representative who plays an important role in the establishment of an organization's policies or is responsible for managing an important aspect of the organization's activities and, in the case of a body corporate, includes a director, its chief executive officer and its chief financial officer.[129]

Both definitions cover broader categories of corporate personnel than the 'directing mind' concept developed at common law.

In the early 1980s academic commentators, policy-makers and judges also began to recognise that imposing corporate liability would not, in itself, necessarily change corporate policy and promote future environmental compliance. This led to calls for the personal liability of directors and other senior officers. Chief Justice Stuart sitting in the Yukon Territorial Court, echoed these sentiments, stating that:

> Fining corporations leaves the upper echelon policy makers relatively unscathed. Fining corporate policy makers reduces somewhat the impotency of levying fines against corporate assets. [Personal liability] is the most effective method of ensuring that persons with the power to shape corporate policy are deterred from either active or passive acquiescence in the development of corporate policies precipitating violations.[130]

In the late 1980s, another northern judge, in sentencing a company for a violation of the Fisheries Act, ordered the directors and Chief Executive Officer of the company to issue a public apology for the company's negligence.[131] In this instance, the order was struck down on appeal, but it was becoming clear that in pursuing corporate compliance, directors and other senior officers should be held accountable for the actions of their company. Over the past 15–20 years, there has been a notable growth in legislative provisions that expressly provide for individual director and officer liability.[132] Section 280 of CEPA, for example, provides that:

> Where a corporation commits an offence under this Act, any officer, director or agent of the corporation who directed, authorized, assented to, acquiesced in or participated in the commission of the offence, is a party to and guilty of the offence, and is liable to the punishment provided for the offence, whether or not the corporation has been prosecuted or convicted.

[128] S 22.1 Criminal Code.

[129] S 2 Criminal Code.

[130] *R v United Keno Mines Ltd* (1980) 10 CELR 43, [40], [48].

[131] *R v Northwest Territories Power Corp* (1980) 10 CELR 43, [40], [48].

[132] Corporate officials have always been liable as parties to a corporate offence on the basis that they aided or abetted the corporation in the commission of the offence,

Despite the introduction of these explicit provisions, great uncertainty still surrounded the imposition of director/officer liability. Did prosecutors have to establish actual influence on the act or omission or was it enough to demonstrate control of relevant aspects of corporate policy and activity? Partly in response to these ambiguities, environmental offences now often include provisions which require directors and other corporate officers to take reasonable care or exercise due diligence. In 1986, a section was added to the Environmental Protection Act (Ontario), for example, requiring directors and officers of a corporation that engages in an activity that may result in the discharge of a contaminant into the natural environment 'to take all reasonable care to prevent the corporation from causing or permitting such unlawful discharge.'[133] Comparable duties are found at federal level in section 280(2) of CEPA. In *R v Bata Industries Ltd*, the first case in which directors were successfully prosecuted for environmental offences under the EPA (Ontario), it was established that the defendant company officers had to prove, on the balance of probabilities, that they took all reasonable care.[134] Although this reversal of the burden of proof was rejected on appeal, the decision of the Ministry of the Environment (Ontario) to prosecute Bata officers undoubtedly raised the stakes and contributed significantly to raising the level of awareness of directors' and officers' responsibilities within companies. The relevant section of the EPA has in fact been amended and as originally decided in Bata, directors and other senior officers now have to prove that they exercised reasonable care.

Interestingly, neither Environment Canada's Compliance and Enforcement Policy for CEPA nor the Ministry of Environment (Ontario) Enforcement Policy makes reference to when it would be appropriate to prosecute a company and/or its directors or other senior officers. Saying that, CEPA enforcement statistics and available information on MoE (Ontario) enforcement suggests that the majority of prosecutions are brought against corporate entities as opposed to individual defendants generally.[135] Despite overwhelming support for director liability, it would seem that very few company officers have been successfully prosecuted.

D Corporate and Director Liability in England and Wales

In UK criminal law generally, vicarious liability and the identification doctrine are used to attach criminal liability to a company. But as in Australia and Canada, where the limitations of these two approaches have been acknowledged and alternative corporate liability models developed, the UK courts have introduced a more purposive approach to statutory interpretation whereby they interpret the

[133] EPA 1990 (Ontario) s 194.

[134] Above n 73.

[135] As stated in Chapter 5, there is a noticeable lack of information on environmental law enforcement in Ontario.

offence to determine whose acts will be attributed to the company.[136] It would appear that this approach has been utilised most flexibly in the environmental context. The case of *NRA v Alfred McAlpine Homes East Ltd*[137] provides a good illustration of the purposive approach. The defendant company was charged with causing pollution of controlled waters under section 85(1) of the Water Resources Act 1991. Those responsible for the pollution were two relatively lowly managers who were clearly not part of the directing mind of the company. Applying the identification doctrine, the company was acquitted in the magistrates' court. On appeal to the High Court, Morland J adopted a purposive construction of the legislation, stating that:

> To make an offence an effective weapon in the defence of environmental protection, a company must by necessary implication be criminally liable for the acts or omissions of its servant or agents during activities being done for the company.[138]

According to Morland J, the purpose of the section 85 offence was to control pollution. He argued that most pollution incidents would be directly caused by employees far removed from the directing minds of a company and therefore, if a company was not liable for the acts and omissions of such employees, then 'the effectiveness of the relevant part of section 85(1) would be lost and the filthiness of rivers increased.'[139] In other words, due to their ability to train, discipline and supervise their staff, companies are best placed to prevent or minimise pollution. Simon Brown LJ approached the case from a slightly different angle, focusing on the much discussed case of *Alphacell v Woodward*[140] and the meaning of the word 'cause'. In his opinion, a company would be liable for pollution providing it resulted from the operations of the company and that third party acts did not break the chain of causation.

Both opinions were endorsed in the case of *Shanks and McEwan (Teeside) v Environment Agency*, when the supervisor of a landfill site failed to complete a waste transfer form when redirecting the waste to a containment bund.[141] The company was held liable in two ways: first, that the 'controlling' minds of the company knew that waste was deposited at its landfill site (in other words, it was not necessary to prove that the 'controlling' minds of the company knew that the waste had been redirected); and second, that the supervisor's knowledge of the incident could be attributed to the company. It would therefore seem that there are few problems in ensuring corporate liability for environmental offences, thus

[136] Lord Hoffman, in the Privy Council case of *Meridian Global Funds Management Asia Ltd v Securities Commission* [1995] 3 WLR 413 framed the question in the following way: 'Whose act (or knowledge, or state of mind) was for this purpose intended to count as the act etc. of the company?' Although the Court of Appeal has declined to consider extending the identification doctrine in this way, there are, as we shall see, some examples of the courts in the UK using this approach in environmental cases.

[137] [1994] Env LR 198.

[138] *Ibid* 211–12.

[139] *Ibid* 213.

[140] Above n 50.

[141] [1997] Env LR 305.

strengthening the deterrent impact of criminal prosecution. This is especially so considering neither of the two approaches relies on proof of mens rea.

Individual 'directorial' liability is also provided for in most if not all environmental statutes. For example, under regulation 32(4) of the Pollution Prevention and Control (England and Wales) Regulations 2000 any director, manager, secretary or other similar officer of the company can be prosecuted personally if the offence is committed with the consent or connivance of, or is attributable to their neglect.[142] The liability applies primarily but not exclusively to directors, although what really matters is whether or not the individual had sufficient responsibilities, in terms of controlling and guiding corporate policy and strategy to amount to the 'controlling mind' of the company.[143]

The Environment Agency's Enforcement and Prosecution Policy gives further guidance as to when it will prosecute companies and/or senior managers, stating that it will usually prosecute the company where the offence has resulted from the company's activities.[144] Action will also be taken against individual officers where it can be shown that 'the offence was committed with their consent, was due to their neglect or they 'turned a blind eye' to the offence or the circumstances leading to it.'[145] From a practical perspective, the Agency's Spotlight Reports give some indication as to the number of companies prosecuted, although it is difficult to compare one year with another due to differences in how the statistics are presented year on year. Generally speaking, the Agency prosecutes more individual defendants (not specifically corporate executives) than it does companies. The majority of companies that are prosecuted are SMEs. Unlike their counterparts in Australia and Canada, the Agency publishes annual statistics on the number of directors or other senior managers prosecuted annually, with the number rising gradually over the past five years, from 11 in 2003 to 29 in 2006.[146]

V Conclusion

There is widespread public support for treating pollution violations as crimes, and indeed in many countries offences are underpinned by comprehensive systems of criminal liability whereby offenders can be brought to account through the criminal process. The imposition of criminal liability (and consequently criminal sanctions) on those who cause environmental harm or carry out activities which

[142] SI 2000/1973. NB, there must be an offence committed by the company.

[143] See *Woodhouse v Walsall MBC* [1994] Env LR 30.

[144] Environment Agency, *Enforcement and Prosecution Policy* (August 2008) para 25, available at: <http://www.environment-agency.gov.uk/static/documents/enforcement-policy.pdf>accessed 7 January 2009.

[145] *Ibid.*

[146] See Environment Agency *Spotlight Reports* available at <http://www.environment-agency.gov.uk> accessed 5 January 2009.

have the potential to cause such harm is arguably the only appropriate response to such antisocial behaviour. With scientific evidence increasingly pointing to the large-scale, irreversible destruction of the local, national and global environment, the criminalisation of polluting behaviour arguably reflects society's moral condemnation of such activities. The use of the criminal law and criminal sanctions is also justified in terms of its deterrent effect on potential offenders and the subsequent prevention of future harm. In summary, the key assumption seems to be that the criminal enforcement of pollution control regulation may lead to improved deterrence, enhanced compliance and thereby a cleaner environment.[147]

In this chapter, I examined how law-makers in Australia, Canada, and England and Wales have modelled offences with a view to maximising the deterrent effect of pollution control prosecutions. The imposition of 'strict liability' and the relative ease of prosecution can assist regulators in achieving the dual goals of compliance and deterrence. From an ethical perspective, all environmental offences should impose 'strict liability'. The potential severity of harm and the impact of damage on future generations necessitates that the highest standard of precaution should be expected of parties who knowingly undertake regulated activities. I argue that whilst there are strong arguments against the imposition of 'strict liability' for environmental crimes, particularly in its application to individuals, sole traders and micro businesses, the harshness of such a liability regime is mitigated by the role that culpability plays in deciding whether or not to prosecute and in the sentencing of offenders. Coupled, in some instances, with the availability of defences such as due diligence, which highlights the need for operators to have policies and procedures, strict liability is appropriate in criminalising polluting behaviour.

In England and Wales, the majority of environmental offences are based on strict liability. In contrast, legislative bodies in Australia and Canada have adopted a more sophisticated model of liability by creating a range of offences, some of which are based on strict liability while others depend on proof of some subjective or objective mens rea. The maximum monetary fine corresponds to the degree of culpability required to satisfy the legislative definition of the crime. So, for example, crimes which require proof of intention carry much higher fines than crimes which are assessed on the basis of a negligence standard. The 'tiering' of offences (and penalties) in this way, has two distinct advantages. As provision is made for crimes committed in the absence of mens rea, regulators are able to utilise, at relatively low cost, the criminal justice procedure to condemn polluting behaviour. But second, where the regulator is able to prove intent, recklessness or negligence, the offender can be prosecuted for a fault-based offence which may have a greater stigmatising effect on the guilty party, therefore increasing the deterrent value of

[147] For an interesting and broad-ranging discussion of criminal environmental law, see R Malcolm, 'Prosecuting for Environmental Crime: Does Crime Pay?' (2002) 14 *Environmental Law and Management* 289.

the penalty. The legislation also recognises that strict liability crimes are less justifiable in the case of individuals, and prescribes different penalty scales for individuals and corporate offenders. In my view, this more nuanced approach to liability undoubtedly has its advantages and should be seriously considered for adoption in England and Wales.

Furthermore, making corporate entities and their directors and senior managers liable for pollution offences also strengthens the enforcement hand of environmental regulators, who are able to target their enforcement action against those most able to ensure future compliance. Legislative bodies and the courts in Australia, Canada, and England and Wales use a range of approaches in attaching liability to companies and their directors and it would seem that regulators and prosecutors see this as important in deterring non-compliance. However, the deterrent impact of directorial liability, in particular, is weakened by two interrelated factors: (a) the problems inherent in attaching liability to corporate executives; and (b) the seemingly infrequent exercise of discretion in favour of prosecuting these individuals.

In the next chapter, I take my analysis of the criminal law one step further, and consider the range of criminal sanctions available to the courts when sentencing offenders. The deterrent impact of these sanctions is clearly at the core of the deterrence framework developed in Chapter 2. Whilst strong arguments can be put forward as to why polluting behaviour should be criminalised, it can, however, be extremely difficult to find a sanction or set of sanctions which will ensure effective deterrence. The deterrent effect of criminal sanctions in the area of environmental regulation is widely disputed, particularly in the case of corporate entities which famously have 'no soul to damn: no body to kick.'[148] It is to these matters to which I now turn.

[148] Taken from the title of Coffee's seminal work on the punishment of corporations: JC Coffee, 'No Soul to Damn: No Body to Kick: An Unscandalized Inquiry into the Problem of Corporate Punishment' (1980) 79 *Michigan Law Review* 386.

7

Judicial Sanctions and Deterrence

I Introduction

HAVING EXAMINED THE nature and scope of criminal liability in pollution control regulation, my analysis advances to consider the range of judicial sanctions (both civil and criminal) which can be utilised by the courts. In other words, it is concerned directly with the nature of the sanction (D) in the deterrence framework developed in Chapter 2. According to Becker's deterrence framework, which provides a valuable analytical tool in the study of law and enforcement, the 'value' of the sanction (along with the likelihood of prosecution) is key in predicting the deterrent impact of the criminal law. The higher the sanction, the more likely it is that, when discounted by the probability of prosecution and conviction, it will outweigh the benefit accruing to the offender from the offence. The chapter focuses on the multiplicity of criminal penalties which can be imposed subsequent to a conviction. With this in mind, a couple of important points need to be made at the outset. First, increasing the expected costs of punishment (ie, D) is merely one way to reduce criminal activities. Criminal activity may also be influenced by, for example, increasing the opportunity costs of crime by reducing unemployment.[1] Second, when considering the expected costs of criminal sanctions, one should remember that the deterrence value in a finding of guilt in the criminal court extends beyond the boundaries of any formal penalty imposed. Some would suggest that the moral stigma associated with criminal conviction is the most efficient punishment of all, particularly in its application to larger corporate entities who have a reputation to damage.[2] A criminal conviction can also prevent the offender from holding an environmental licence or permit in the future. This in itself performs an important deterrence function. In addition, and unlike civil or administrative remedies, a criminal conviction will normally result in the imposition of a punitive sanction which is designed primarily to deter the offender (and others) from committing offences in the future. The theories of general and specific deterrence are in fact viewed as one of the rationales behind

[1] R Posner, *Economic Analysis of Law*, 7th edn (Aspen Publishers, USA, 2007) 19–22.

[2] See, eg, J Lott and JM Karpoff, 'The Reputational Penalty Firms Bear for Committing Fraud' (1993) 36 *Journal of Law and Economics* 757.

the sentencing of offenders. Unsurprisingly then, deterrence rationales are at the core of my analysis.

The chapter reviews, in Sections III and IV, the conventional sanctions prescribed for pollution control offences, namely fines, custodial sentences, community penalties and other non-custodial sentencing alternatives. Monetary fines, considered in Section III, are viewed by the Sentencing Advisory Panel in England and Wales as the 'starting point' for dealing with environmental offences.[3] They are flexible and can, in theory, reflect the particulars of the offence such as the harm caused and the previous history of the offender. Imprisonment and other non-custodial sanctions, discussed in Section IV, are also available, although their use is relatively rare and they are obviously not appropriate where the offence is committed by a corporation. The deterrent impact of these 'traditional' criminal sanctions is clear. From an economic perspective, fines are cost-neutral and represent a straightforward transfer of money from the individual to the state. Whilst there are, of course, administrative costs associated with the collection of fines, they are relatively small compared with, in particular, imprisonment which imposes huge costs on society. Therefore, presuming that all fines are collected, they undoubtedly represent from an economics perspective the most cost-effective 'traditional' criminal sanction.

But the use of fines to deter undesirable corporate behaviour in particular has been called into question for some time. In response, alternative or creative sentencing options, primarily the use of non-fine measures, have been advanced as a response to the perceived failure of traditional criminal sanctions to deter offenders and promote compliance. Sections V and VI focus on these creative sentencing options, their strengths and weaknesses and their use by the courts in sentencing polluters. Sanctions such as adverse publicity orders and environmental project orders are prevalent in pollution control regulation in Australia and Canada and there have been calls for their introduction in England and Wales. The direct financial impact of these creative sentencing options would clearly have a deterrent effect on both individual and corporate offending. But, importantly, they may also have additional benefits, many of which reflect criminal sentencing rationales other than deterrence. For example, environmental audit orders (requiring the revision of policies and practices) reflect rehabilitative goals. In contrast, environmental service orders (requiring the offender to carry out an environmental project) arguably give effect to rationales based on restorative justice. Alternative or creative sentencing has particular strengths when applied to corporations. First, it can reflect the important non-monetary motives of companies such as the desire for power and prestige. And second, although it is not as obviously relevant in the context of corporations, commentators such as Brent Fisse have argued that alternative sentencing goals are, in fact, subgoals of deterrence in corporate crime.[4] Many environmental offences occur as a result of, for example, substan-

[3] Sentencing Advisory Panel, *Environmental Offences: The Panel's Advice to the Court of Appeal* para 14.

[4] B Fisse, 'Reconstructing Corporate Criminal Law: Deterrence, Retribution, Fault and Sanctions' (1982) 56 *South Californian Law Review* 1141.

dard operational procedures and poor internal disciplinary control. Where a corporate offender is punished (or threatened with punishment), then the message conveyed by the punishment is arguably catalytic as well as inhibitory. In other words, the message is 'refrain from committing that offense and take such steps as are necessary organizationally to guard against repetition.'[5] All in all, creative sentencing options such as those available in Australia and Canada undoubtedly strengthen the deterrence signal of the criminal law. Their introduction should seriously be considered here in the UK. The chapter concludes by examining, albeit briefly, the availability and use of civil enforcement proceedings in pollution control regulation.

Before discussing, in detail, criminal sanctions and their deterrent effect, Section II begins by briefly considering deterrence, and other rationales of sentencing such as rehabilitation and retribution. Many of the sentencing options available to the courts in pollution cases reflect one or more of these rationales and so it is important to appreciate alternative penological approaches to sentencing. Section II also provides an overview of sentencing considerations, more specifically the aggravating and mitigating factors that courts take into account in determining an appropriate sentence. Courts in Australia, Canada, and England and Wales have regard to a common set of factors, and so guidance available to courts in England and Wales is used as an example.

II Sentencing and Discretion

A Introduction

Sentencing is a vital part of the criminal justice system. But determining the appropriate punishment for criminal conduct is a difficult task. Judges must have considerable discretion in sentencing offenders, thus enabling them to tailor their sentence to both the particular penal philosophy underpinning the type of offence, and the subjective and objective characteristics of the offence and the offender. But this discretion is not unchecked, and in recent decades the degree of structuring has arguably increased. There are certain obvious ways in which discretion in sentencing is curtailed. First, judges can only use penalties which are legally available to them. And second, there will be rules governing the use of available sentencing. So legislation may, for example, prescribe a maximum fine or may state that certain sentencing powers are not available for offenders under the age of 16. Discretion may also be structured by both common law and sentencing procedure statutes which enunciate the purposes for which a court can impose a sentence on an offender. Of course, the existence of discretion does not necessarily mean that a criminal court will attempt to tailor the sanction to what is required by the

[5] *Ibid* 1160.

economic deterrence framework (as described in Chapter 2). When sentencing polluting offenders, judges may, in fact be pursuing other objectives, many of which reflect different penological principles used to justify punishment. These include deterrence (both general and specific), rehabilitation, protection of the public (incapacitation) retribution (punishment), and restoration and reparation.[6] Before considering in detail the sentencing options available to the criminal courts in pollution cases, it is useful to examine briefly these penological principles as well as other sentencing considerations utilised by the courts.[7]

B Penological Principles

Deterrence, rehabilitation and protection of the community, clearly have a preventative aim. Deterrence is, in fact 'one of the main purposes of the criminal law.'[8] It is concerned with the extent to which different sentencing options prevent: (a) the offender from re-offending (specific deterrence); and (b) members of the public (or in our case, the regulated community) from committing a similar kind of offence (general deterrence). Under a sentencing system based on specific deterrence, the courts would have to have detailed information on the character of the offender and her previous criminal record. This would enable the courts to set a sentence that would deter that particular offender. In such cases, the main determinate of sentencing would have to be the likelihood of the offender reoffending, as opposed to the seriousness of the offence. General deterrence, on the other hand, is based on utilitarian rationales, and justifies punishment where the benefits (in terms of general deterrence) outweigh the costs (to the offender) of punishment. Although there is some doubt surrounding the capacity of the criminal law to deter street criminals, there would appear to be some support for the fact that general deterrence does work in relation to white-collar crime, including environmental cases.[9] General deterrence can be explained and discussed using the deterrence framework in Chapter 2, which predicts when criminal prosecution and criminal sanctions will deter an offender, as measured against the benefit or utility the offender derives from committing the offence. Rehabilitation as a goal

[6] See eg Criminal Justice Act 2003 (UK) s 142 and Crimes (Sentencing Procedure) Act 1999 (NSW) s 3A. Also see B Preston, 'Principled Sentencing for Environmental Offences—Part I: Purposes of Sentencing' (2007) 31 *Criminal Law Journal* 91 for a detailed discussion of penal philosophies in New South Wales, Australia.

[7] For details on sentencing see S Easton and C Piper, *Sentencing and Punishment—The Quest for Justice* (Oxford, OUP, 2005).

[8] AI Ogus, *Costs and Cautionary Tales: Economic Insights for the Law* (Oxford, Hart Publishing, 2006) 119.

[9] See eg JD Silberman, 'Does Environmental Deterrence Work? Evidence and Experience Say Yes, But We Need to Understand How and Why' (2000) 30 *Environmental Law Reporter* 105, 123, cited in Preston above n 6, 94. Also see, eg D Thornton, NA Gunningham and RA Kagan, 'General Deterrence and Corporate Environmental Behavior' (2005) 27 *Law and Policy* 262, in which a survey of 233 firms in several industries in the US revealed weak support for the general deterrence hypothesis. They also found that general deterrence had a cumulative effect on companies, in that it reassured them that compliance was worthwhile.

of sentencing also has a preventative aim, and seeks to reform offenders by forcing a change of behaviour. A rehabilitative sanction in a pollution case may, for example, educate and train the offender in pollution abatement procedures and techniques. The third principle that serves a preventative function is protection of the public or incapacitation.[10] This principle most obviously manifests itself in the availability of custodial sentences; by imprisoning offenders, they are removed of their ability to commit further offences. Other penalties may also have community protection at their core. For instance, one direct or indirect consequence of successfully prosecuting polluters is that the offender may be prevented from holding an environmental licence or permit in the future. Incapacitation and other sentencing options designed to protect the public inevitably put the needs of the community ahead of the welfare of the offender. According to Preston:

> This prioritisation is relevant in environmental cases where there is an evident need to protect the community from environmental crime. Environmental offences concern the public welfare. They involve a shift of emphasis from the protection of individual interests to the protection of public and social interests.[11]

Criminal sanctions designed to rehabilitate and/or incarcerate offenders most often create a non-financial disutility to the actor. What is questionable, however, is the extent to which they impose sufficient disutility to deter the offender from committing an offence. Imprisonment imposes a severe non-financial cost on the offender and so would normally constitute an effective deterrence. Community service and other rehabilitative sanctions may not.

Retribution as a sentencing rationale differs significantly from the preventative principles discussed above. Underlying retributive rationales is the quest for justice; crime is immoral and punishment for crime is morally right. Offenders should therefore be punished according to what they deserve ('just deserts').[12] Although in recent years, retribution has possibly taken precedence over deterrence, Ogus in his discussion of behavioural incentives argues that the two approaches may, in fact, be compatible.[13] For example, in determining the seriousness of the offence (an important element of retributist theories), the courts will take into account inter alia the culpability of the offender. Whilst this is normally justified by reference to moral considerations, it can also serve an economic function for an offender who commits an offence intentionally or deliberately is likely to derive a higher benefit or utility from the offence. This therefore means that using the deterrence framework, a larger sanction would be needed to deter the offender.[14]

[10] Incapacitation is not necessarily incompatible with rehabilitation as a sentencing rationale. In some jurisdictions, including the UK, there is a strong commitment to the use of offending behaviour programmes which are designed to rehabilitate offenders whilst in prison.

[11] Preston above n 6, 97.

[12] For a summary of retribution as a principle of sentencing see Preston *Ibid* 92–4.

[13] Above n 8, 119–21.

[14] See further R Cooter, 'Prices and Sanctions' (1984) 84 *Columbia Law Review* 1523, 1537.

The final rationale of sentencing is restoration and reparation. This recognises the rights and needs of the victims of crime.[15] Over the past four decades, restorative justice, which was never traditionally associated with punishment, has appeared in the literature on crime and sentencing. Braithwaite, a leading exponent of restorative justice, believes that it has two important dimensions.[16] The first deals with process; that the victim (and possibly others involved in the harm) should in some way participate in the criminal justice system by being given, for example, the right to present a victim impact statement to the court when sentencing takes place. The second reflects values such as reintegration and forgiveness, both of which, according to restorative justice theories, 'trump' other values such as retribution and punishment. So, an offender may be required to apologise to the victim, in some way repair the harm that has been caused or even pay the costs and expenses incurred by a public authority in investigating the offence. From our perspective, it is interesting to consider who the 'victims' of environmental crime are. This question can be looked at from both an anthropocentric and ecocentric perspective. First, it could be argued that natural resources such as air, water and land are for the use of the general public rather than a particular group of people so the victim is usually the public at large. Second, and taking this one step further, under the principle of inter-generational equity, future generations may also be classed as victims of environmental crime.[17] And third, is it not logical to conclude that the actual victim of environmental crime is 'the environment' itself, including flora and fauna? Whatever the approach, sanctions reflecting restorative justice values may create both financial disutility (eg, compensation and reparation of harm) and non-financial disutility (eg, public apology) to an offender.

C Sentencing Guidelines

In addition to the penological approaches described above, guidelines may help judges apply the sentencing framework as consistently as possible. Sentencing considerations established for pollution offences in many jurisdictions including Australia,[18] Canada[19] and the UK identify a range of common aggravating and mitigating factors that should be taken into account in sentencing offenders. Take the UK as an example.[20] Specific guidelines have been produced for magistrates on

[15] For a general summary, see Easton and Piper above n 7 at ch 6.

[16] See J Braithwaite, *Regulation, Crime, Freedom* (Aldershot, Ashgate Publishing, 2000) and J Braithwaite, *Restorative Justice and Responsive Regulation* (Oxford, OUP, 2002).

[17] Above n 6, 99.

[18] See, eg, in New South Wales, PEOA 1997 (NSW) s 241. A detailed appraisal of sentencing considerations is provided by Preston. See B Preston, 'Principled Sentencing for Environmental Offences—Part 2: Sentencing Considerations and Options' (2007) 31 *Criminal Law Journal* 142.

[19] See, eg, in Ontario, s 188.1 EPA 1990 (Ontario) and generally *R v United Keno Hill Mines Ltd* (1980) 10 CELR 43.

[20] For detail see N Parpworth, 'Environmental Offences: The Need for Sentencing Guidelines in the Crown Court' [2008] *Journal of Plannning and Environment Law* 18.

the sentencing options available to them for environmental and health and safety offences.[21] More generally, in 2001, the Sentencing Advisory Panel, empowered to make proposals to the Court of Appeal for new guidelines, examined a number of environmental offences and advised the Court of Appeal to provide sentencing guidelines.[22] The Court of Appeal declined the invitation, preferring to rely on the factors it identified in the health and safety case of *R v F Howe and Son (Engineers) Ltd* many of which are applicable in the environmental field.[23] The court referred to aggravating factors such as the degree to which the defendant fell short of his duty, the degree of risk and extent of danger created, a failure to heed warnings, the extent to which the defendant deliberately ran the risk for profit and the nature of injury or damage. Mitigating factors included the prompt admission of responsibility, the taking of steps to remedy the problem, and the previous good record of the offender. Three years later, in the environmental case of *R v Yorkshire Water Services Ltd*, the Court of Appeal adopted the guidelines in Howe but also recognised that additional factors would apply in environmental cases, including the spatial and temporal ambit of the damage caused by the offence.[24] Similar aggravating and mitigating considerations are reflected in statutory sentencing considerations in Australian and Canadian pollution control legislation.

III Fines and Deterrence

A Introduction

The overwhelming majority of pollution cases result in the imposition of a monetary fine. As such, they form an extremely important part of the criminal courts' arsenal in sentencing polluters. But this is not to suggest that monetary fines are best able to deter offenders. Their deterrence impact is limited by inter alia the fact

[21] Magistrates' Association, *Fining of Companies for Environmental and Health and Safety Offences* (May 2001) and Magistrates' Association and Environmental Law Foundation, *Costing the Earth—Guidance for Sentencers* (October 2003).

[22] Above n 3. The Sentencing Advisory Panel was established by the Crime and Disorder Act 1998 ss 80, 81. Following the creation of the Sentencing Guidelines Council pursuant to Criminal Justice Act 2003 s 167, the Sentencing Advisory Panel continues to propose to the Council that specific sentencing guidelines should be reviewed.

[23] [1999] 2 All ER 249

[24] [2002] Env LR 18 cited in Parpworth above n 20, 26–7. The extent to which these sentencing guidelines have assisted the lower criminal courts in sentencing environmental offenders in the UK is debatable. This is particularly apparent in fines handed down by the Crown Court in environmental cases, some of which have been reduced considerably on appeal. Parpworth (at 28) argues that these reductions are worrying, not because they suggest that the Court of Appeal is too lenient, rather that the Crown Court is unable to arrive at a just and appropriate penalty. He states that '[T]he Court of Appeal has had several opportunities to lay down clear sentencing guidelines but instead, it has chosen to borrow guidelines which it has itself framed in relation to a category of offence, health and safety, where liability is not strict and where the commission of an offence may often involve human fatality.'

that a fine cannot be effective if the recipient cannot afford to pay it. This may go some way to explaining the seemingly low levels of fines handed down by the courts in Australia, Canada, and England and Wales. This section begins by summarising the provision made for fines in pollution control regulation in Australia, Canada, and England and Wales. It then considers in Sections C and D the strengths and weaknesses of fines from both a theoretical and practical perspective. It finishes by examining two ways in which the deterrent impact of fines, as a criminal sanction, could be strengthened, focusing on the use of equity fines.

B Legislative Provisions: An Overview

In most jurisdictions where non-compliance with pollution control law is a criminal offence, legislative provision is made for the imposition of a fine by the courts. This section provides an overview of the key statutory provisions relating to criminal fines.

i Australia

In Australia, regulatory provisions will usually prescribe the maximum fine that can be imposed for a particular offence. As noted in Chapter 6,[25] there are different penalty schemes for individuals and corporations and it is common for higher maximum penalties to apply to offences requiring an element of mens rea, arguably reflecting the greater degree of moral culpability in such cases.[26] For example, in Victoria, a range of financial penalties can be imposed for offences committed under the EPA 1970. Offences of aggravated or intentional pollution are punished more severely than those committed negligently or otherwise and an offence committed by a corporate entity will usually be more severely punished than if the same offence were committed by an individual. So for example, the maximum fine for a corporate body that commits an aggravated (level 1) pollution offence is $1,074,300 whereas for an individual it is $268,575 and/or seven years' imprisonment.

In some states such as Victoria and NSW, maximum fines have in fact been dramatically increased in recent years.[27] Take, for example, the maximum penalties now provided for in the PEOA 1997 (NSW).[28] As illustrated in Table VI below, the new maxima are twice if not five times higher than those originally provided for.

[25] See s III A.

[26] For a discussion of the merits of strict liability in criminal law see ch 6 s II B.

[27] Environment Protection (Enforcement and Penalties) Act 2000 (Victoria). Amendments to the PEOA 1997 (NSW) are discussed below.

[28] The Act was amended by the Protection of the Environment Operations Amendment Act 2005, the provisions of which came into force in May 2006.

Table VI: Offences and Penalties under the Protection of the Environment Operations Act 1997 (NSW)

Offences	Penalty Range
Tier 1 (Part 5.2 of the Act)[1]	Up to $5 million *($1 million)** for corporate body
Offence committed wilfully	Up to £1 million *($250,000)* and/or 7 years imprisonment for an individual
Offence committed negligently	Up to $2 million *($1 million)* for corporate body
	Up to $500,000 *($250,000)* and/or 4 years imprisonment for an individual
Tier 2 (all other offences)	Up to $1 million *($250,000)* for corporate body (and up to $120,000 each day for continuing offence)
	Up to $250,000 *($125,000)* for individual (and up to $60,000 each day for continuing offence)

* Italicised figures represent the maximum penalties prior to the passing of the Protection of the Environment Operations (Amendment) Act 2005.

[1] Tier 1 offences include the disposal of waste in a manner that harms or is likely to harm the environment (s 115) and causing a substance to leak, spill etc. in a manner that harms or is likely to harm the environment (s 116). See further Ch 6 Sec III A(iv).

ii Canada

As is the case in Australia, Canadian provinces such as Ontario have a different penalty structure for companies and individuals, with corporations facing higher fines if convicted in the criminal courts. Offences are also 'tiered' according to the type of offence committed. So, offences which create a risk of an adverse environmental impact attract potentially higher fines than a general breach of the regulations. In some provinces, including Ontario, penalties are calculated on a daily basis and, interestingly, higher penalties can be imposed in the case of subsequent convictions.[29] As in Australia, the level of fines authorised by statute has been increased significantly since the 1980s. Most recently, in Ontario, the Environmental Enforcement Statute Law Amendment Act 2005 (Bill 133) raised the penalties substantially (see Table VII below). It also introduced new minimum penalties[30], increased the categories of offences for which the more serious penalties can be imposed and included a new 'second conviction' tier of penalties.

iii England and Wales

The fines provided for in pollution control regulation relevant to England and Wales are much less nuanced. The legislation provides for no minimal levels, there

[29] EPA 1990 (Ontario) s 187.1(b).

[30] But under s 59 of the Provincial Offences Act, the court can undercut the minimum penalties.

Table VII—Examples of Penalties (Day Fines) under the Environmental Protection Act 1990 (Ontario) section 187

Description	Individual	Corporation		
	First Offence	Subsequent	First Offence	Subsequent
Level 1—General Contravention of Act or Regs	Up to $50,000 (*$20,000*)	Up to $100,000 (*$50,000*); max one year imprisonment	Up to $250,000 (*$100,000*)	Up to $500,000 (*$200,000*)
Level 2—Risk of adverse effect, providing false information, violation of an order, certificate of approval licence or permit requirement	Min of $5,000 up to $4 million (*up to $50,000*)	On second conviction, minimum of $10,000 up to $6 million; max five years less one day imprisonment. On subsequent conviction minimum of $20,000 up to $6 million; max five years less one day imprisonment (*up to $100,000; max one year imprisonment*)	Min of $25,000 up to $6 million (*up to $250,000*)	On second conviction, min of $50,000 up to $10 million. On subsequent conviction min of $100,000 up to $10 million (*up to $500,000*)

* Italicised figures represent the maximum penalties prior to the passing of the Environmental Enforcement Statute Law Amendment Act 2005

is no separate scale for corporations and individuals, and offences are generally not tiered according to mens rea or environmental harm criteria. The maximum penalties do, however, depend on the court in which the offence is tried. Most environmental offences are triable 'either way', that is they can be dispatched summarily in a magistrates' court, or by a jury in a Crown Court. The general maximum fine for any one offence is £5,000 but in most cases, breach of environmental law attracts a higher maximum fine, normally £20,000 although recent legislative amendments have increased the statutory maxima for certain offences.[31] Magistrates cannot award a fine and/or imprisonment above the maximum provided for, so if a person has been convicted summarily of an offence triable either way he may sometimes be committed to the Crown Court for sentencing. On

[31] A maximum fine of £20,000 is available to Magistrates under, eg, Water Resources Act 1991 s 85, which makes it an offence to cause or knowingly permit any poisonous, noxious or polluting matter or any solid waste matter to enter any controlled waters. Clean Neighbourhoods and Environment Act 2005 s 41 increased the maximum fine in a magistrates' court to £50,000 for waste offences.

committal for sentence to the Crown Court, the court will inquire into the circumstances of the case in the same way as if the defendant had been convicted there. There is generally no limit to the level of fine that can imposed by the Crown Court.

C The Deterrent Impact of Fines: A Theoretical Perspective

From a deterrence perspective, monetary fines would appear to have a number of advantages over other criminal penalties. First, fines are cost-neutral: the convicted criminal is fined £1,000, and the court system collects £1,000. The money may be used to run the court system itself, it may be used to compensate the victims of crime or it may generate an increase in enforcement activity out in the field. What one can be sure of is that the money will be transferred, either directly or indirectly, to someone other than the offender. Furthermore, presuming they are paid promptly, fines are relatively easy to administer; generally speaking, the smaller the fine, the more likely it is that the criminal will be able to pay it, and therefore the less costly it is to collect. So, in short, fines minimise the costs of sentence implementation to society and may, in effect, directly transfer the economic benefit obtained by the offender through the offence (or the resulting costs imposed on the environment) from the offender to the public. As noted by Campbell,[32] this economics approach, which justifies punishment on the basis that it internalises externalities such as pollution, has been judicially endorsed:

> The economic reason why society may not in the absence of regulation strike a proper balance between economic output and environmental quality is that the costs of pollution are borne not by polluters but by somebody else. As a result, the 'external' costs will not, in general, be taken fully into account by those who cause pollution. In so far as pollution costs are not borne by those who cause pollution, or by the purchasers of their products, some part of the total benefits resulting from economic activity in that community is wrongly distributed away from the victims of pollution to other groups in society. In order to correct this market failure, the government must intervene to impose financial costs or penalties which bring the external costs back to the polluter.[33]

A second advantage of fines is that they are extremely flexible and can easily respond to the heterogeneity of the offence, the circumstances of the offender and, most importantly from a deterrence perspective, the utility or benefit the offender derives from committing the offence. From a retributive perspective, they are also relatively easy to fix in terms of proportionality, there being, as noted by Easton and Piper, 'a very detailed tariff of pounds and pence'.[34]

Despite the fact that fines are most easily tailored to the deterrence framework, there are serious questions surrounding the extent to which they can actually deter

[32] G Scott Campbell, 'Fostering a Compliance Culture Through Creative Sentencing for Environmental Offences' (2004) 9 *Canadian Criminal Law Review* 1, 5–6.

[33] *Machinery Movers Ltd v Auckland Regional Council* [1994] 1 NZLR 492, 502 (HC).

[34] Above n 7, 224.

non-compliance. The discussion that follows is particularly relevant when fines are used as a corporate as opposed to an individual sanction. This is partly so because where an individual is unable to provide monetary redress, the criminal justice system has historically provided a range of alternative sanctions, the most serious being incarceration. But this is obviously not an option where the offender is a corporate entity.[35]

Let us begin by considering why fines may not convey the message that serious corporate offences are socially intolerable. First, 'monetary penalties may convey the impression that offences are purchasable commodities or a cost of doing business'[36]. In other words, it may be more economical to pay a fine than it is to address poor environmental performance, especially in larger companies where a fine of $10,000 may 'affect an individual but hardly register with [the] corporation'.[37] For example, in 2001, Esso Australia Pty Ltd was found guilty of 11 offences under the Occupational Health and Safety Act 1985 (Victoria), as a result of which two people died and seven were seriously injured.[38] The company was fined a staggering $2 million, but, considering that the company earned, at that time, approximately $2 million per day from its Bass Strait operations alone, the deterrence value of the fine was, to say the least, limited. So, in order for a fine to be effective, it must be sufficiently large relative to company turnover. Second, the maximum fine that can be imposed is constrained by the wealth of the offender. This is described as the 'deterrence trap' and although it is obviously less of a problem where the offender is an individual as they can be imprisoned,[39] for the corporation 'which has no body to incarcerate, this wealth boundary seems an absolute limit on the reach of deterrent threats directed at it.'[40] Third, even if a large fine is imposed on a company, it may force the company into liquidation. The court may then be faced with either putting the company out of business or imposing a fine which in no way reflects the seriousness of the offence.[41]

Fines are also commonly criticised on the basis of their spillover effect on 'innocent' third parties: 'the corporation catches a cold, someone else sneezes'.[42] It has been argued[43] that, through reduced dividends, wage or job losses and price increases, it is the shareholders, the employees or the consumers who suffer

[35] See generally, B Fisse, 'Sentencing Options against Corporations' (1990) 1 *Criminal Law Forum* 211; JC Coffee, 'No Soul to Damn, No Body to Kick: An Unscandalised Inquiry into the Problem of Corporate Punishment' (1980) 79 *Michigan Law Review* 386.

[36] Australian Law Reform Commission, *Securing Compliance: Civil and Administrative Penalties in Australian Federal Regulation*, (April 2002) Discussion Paper 65, 89.

[37] *Ibid* 89.

[38] *Director of Public Prosecutions v Esso Australia Pty Ltd* [2001] VSC 263.

[39] Although, as we shall see, courts in both jurisdictions are extremely cautious in their use of custodial sentences.

[40] Coffee above n 35, 390.

[41] Other evasion techniques include the use of incorporated subsidiaries and asset stripping.

[42] Above n 35 (Coffee), 401.

[43] See, eg, S Kadish, 'Some Observations on the Use of Criminal Sanctions in Enforcing Economic Regulations' (1963) 30 *University of Chicago Law Review* 423 and JC Coffee and M Jefferson, 'Corporate Criminal liability: The Problem of Sanctions' (2001) 65 *Journal of Criminal Law* 235.

financially from a corporate fine, when in fact they should impact on responsible officers within the corporation.[44] In the long term, they may not change corporate behaviour and may provide no incentive for the company to identify responsible individuals within the firm. Self-discipline within a firm is in many respects, unrealistic and coupled with the fact that, as we have seen, the prosecution of corporate personnel is extremely problematic, this concern with fines is potentially very limiting.[45] Furthermore, fines do not affect the non-financial values that enter into organisation decision-making. Fisse comments that:

> Managerial motivation, like human motivation in general, is not confined to the satisfaction of monetary goals but includes the urge for power, the desire for prestige, the creative urge, and the need for security.[46]

As fines (and other monetary sanctions) only touch upon these non-financial motivations, they only have a limited capacity to sanction.

Finally, as discussed earlier fines would appear to be cost-neutral and cheap to administer. However, this does not always ring true. From a practical perspective the non-payment of fines is a serious problem in many jurisdictions including the UK. For example, the latest National Audit Office's report on Fines Collection revealed that although the payment rate of fines is approximately 80 per cent, 59 per cent of the fines examined required enforcement action of some kind.[47] Although the fees incurred in collecting fines can be added to the fine and thereby extracted from the offender, problems relating to non-payment can dramatically increase the costs associated with the imposition of fines and lends some support to the conclusion that perhaps fines are not as cost-effective in practice as in theory. So, all in all, the deterrence impact of fines is limited and they may not be as cost-effective as they would appear.

In addition to limitations in terms of deterrence, fines are also not necessarily effective in delivering other goals of sentencing such as retribution and rehabilitation. As noted by Braithwaite, the 'retribution trap' is undoubtedly problematic:

> [Can] we imagine any penalty short of revoking the corporation's right to sell drugs which would be commensurate to the harm caused by the fraud and deceit of a thalidomide disaster? Given what we know about how disapproving the community feels toward corporate crime, there may be many situations where the deserved monetary or other punishment bankrupts the company. The community then cuts off its nose to spite its face.[48]

[44] Similar arguments apply to the sentencing of company directors or managers.

[45] For details on 'directorial liability' see ch 6 s IV.

[46] Above n 35 (Fisse), 219

[47] National Audit Office, *Department for Constitutional Affairs: Fines Collection* HC (2005–06, 25 May 2006) 1049. The Department could find no detailed breakdown of why 59% of fines require enforcement action. In its view, the main reason was that fines were set too high due to a lack of information relating to the offender's ability to pay. See further House of Commons Committee of Public Accounts, *Fines Collection* HC (10th Report of 2006–07, January 2007) 245, and the resulting National Audit Office Guide, *Improving fines collection through better information on defendants' ability to pay.*

[48] J Braithwaite, *Challenging Just Deserts: Punishing White-Collar Criminals* (1982) 73 *Journal of Criminal Law and Criminology* 723, 757.

Fines may also not compel a company to review its internal procedures and in some instances, a company may calculate that the costs associated with improving work systems far outweigh the costs of incurring further fines.[49] This is a particular problem in the case of pollution control offences, a large number of which are caused by inadequate management practices. These rehabilitative limitations are illustrated by a now somewhat dated study which examines the impact of prosecutions and fines under the Australian Trade Practices Act. The study concluded that whilst significant organisational improvement occurred in at least 60 per cent of companies who were prosecuted under the Act, in 40 per cent of cases the companies failed to carry out any organisational reform at all.[50]

D Fines and Deterrence: A Practical Perspective

The theoretical limitations of fines are not helped by what occurs in practice. Under the deterrence framework, the expected punishment is calculated by discounting the expected sanction (D) by the probability of prosecution (c). Chapter 4 evaluated the prosecution policies of environmental regulators in Australia, Canada, and England and Wales and concluded that they generally adopt a cautious approach to prosecution. This, in itself, reduces the deterrent effect of the criminal law. If we add to this the fact that the fines imposed by the Australian, Canadian and English courts for pollution offences are also low, and the deterrent effect of the criminal law and criminal fines, as determined by the Becker hypothesis, is even more limited.

In all three jurisdictions, average fines are considerably lower than the maximum penalties available. This is not to suggest that courts should always award fines near the maximum. Rather, that the deferent effect of these penalties will be less meaningful for all but the smallest of businesses. It should be stated at the outset that data on fines is difficult to come by and gaps exist in the information that is available. This makes comparisons and conclusions of fine patterns difficult, both within and between jurisdictions.[51]

Starting with Australia, fines for breaches of Australian pollution control regulation are too low and the courts are rarely using their sentencing powers to the full.[52] Where prosecution has been used in non-adversarial States, the courts have gener-

[49] Mandatory environmental audits address this particular problem. See further s VI E below.

[50] A Hopkins, *The Impact of Prosecutions under the Trade Practices Act* (Canberra, Australian Institute of Criminology, 1978) cited in Fisse above n 35, 225–6.

[51] See further R Kimble, 'Penalties in Regulatory Crime' (2005) 17 *Environmental Law and Management* 169, 173 for a summary of the data that is needed if any valuable conclusions are to be reached on the extent to which (a) fines are high enough and (b) they deter environmental offenders.

[52] For statistics on fines and prosecutions in Victoria and NSW from 1990–2000 see M Hain and C Cocklin, 'The Effectiveness of the Courts in Achieving the Goals of Environment Protection Legislation' (2001) 18 *Environmental and Planning Law Journal* 319, 333–4.

ally awarded very small fines.[53] In the two most adversarial states of NSW and Victoria, statistics paint a slightly different picture. In NSW, fines for EPA prosecutions would appear to be increasing with the average fine rising from approximately $4,718 in 2004–05 to $8,213 in 2005–06 and $22,765 in 2006–07.[54] The dramatic increase in fines for 2006–07 is, according to the DECC, due the fact that the department is focusing on more serious breaches of legislation.[55] In Victoria, the average fine has also increased, albeit more gradually, from $4,727 in 2004–05 to $5,862 in 2005–06 and $5,653 in 2006–07.[56] The lower average in Victoria, compared with NSW, could be partially explained by the fact that the courts utilise their alternative sentencing powers more frequently than in NSW: many of the 'major prosecutions' completed from 2004–07 resulted in the imposition of an environmental project order and an adverse publicity order.[57] Despite the fact that fines are generally going up, calls for further increases have come from a variety of quarters. For example, the Supreme Court of South Australia described a $100 fine handed down by magistrates to a defendant found guilty of unlawfully clearing native vegetation, as 'manifestly inadequate'.[58] The fine was raised to $17,000 on appeal. There is also a concern in academic circles that the 'penalties imposed are generally too low to stimulate extensive and significant environmental improvements to operation and production processes'[59] and that fines are treated as 'simply another form of licensing fee to be factored into their accounts'.[60] The environmental regulators themselves are also concerned that fines do not reflect the seriousness of the offences. This, it has been argued, goes some way to explaining why prosecution levels are relatively low, particularly in the less adversarial states.[61]

In Canada, from the early 1980s through to the mid 1990s, there was actually an unprecedented increase in the level of fines imposed on offenders. According to Diane Saxe, writing in 1989:

> Fines are going up dramatically in environmental cases. Courts around the country, reflecting increasing public interest in protection of the environment, have been imposing previously unheard of penalties. Corporations, individuals and municipalities have all been affected.[62]

[53] J Norberry, 'Australian Pollution Laws: Offences, Penalties and Regulatory Agencies' in AIC Conference Proceedings No 26, *Environmental Crime: Proceedings of a Conference held 1–3 September 1993*, 10.

[54] The DECC (NSW) does not provide data on the average fine. The figures quoted are calculated using information on the number of completed EPA prosecutions and the total value of fines for those prosecutions.

[55] DECC, *Annual Report 2006–07*, 37.

[56] The average fine was calculated using the list of 'major prosecutions' in the EPA's Annual Reports.

[57] See further Table X below.

[58] *Piva v Maynard* [2000] SASC 349, [27].

[59] Hain and Cocklin above n 52, 332.

[60] Cited in Hain and Cocklin *Ibid* 332.

[61] Above n 53.

[62] D Saxe, 'Fines go up Dramatically in Environmental Cases' (1989) 3 *Canadian Environmental Law Reports* (NS) 104, 104.

But due to budgetary constraints, this rise did not continue and in 1996, there was a quite dramatic reduction in prosecutions and fines.[63] Although the total fines collected by the Ministry are now going up, so too are the number of prosecutions. Without specific data on average fines, it is difficult to say whether average fines are going up or down. However, it would seem from correspondence with Ministry officials that average fines are gradually increasing. Again, statistics relating to fines secured by Environment Canada are few and far between, although from those that are available, average fines appear to be going up. However, there remains a concern that fines are still not high enough. From conversations with enforcement officers in the MoE (Ontario) it is clear that fines are still deemed to be too low, with one prosecutor suggesting that the courts have created a distinction between regulatory offences and crimes in the Criminal Code and are subsequently not taking environmental offences seriously enough.[64] Representatives of Environment Canada were of the opinion that whilst the level of fines is increasing, they are still not high enough.[65]

There are similar concerns about the level of fines imposed by the criminal courts in England and Wales. Over the past five years, numerous reviews, reports and government consultations have considered inter alia the effectiveness of fines in sentencing environmental offenders. In February 2005, for example, the Environmental Audit Committee in its report on corporate environmental crime, described fines as verging on the 'derisory in the context of the seriousness of the crime'.[66] Also according to the Macrory Review on Regulatory Justice, the profits gained by offenders are not reflected in the size of the fine which, in effect gives 'businesses an incentive to continue to fail to comply in return for profit'.[67] And most recently, a DEFRA review of the enforcement of environmental regulation concluded that criminal fines are too low in that:

> [T]hey fail to achieve certain key purposes, in particular remediation of damage, restitution to affected communities and removing cost savings or economic gain from non-compliance.[68]

Several explanations have been put forward as to why fines for environmental offences in England and Wales are so low. Even though magistrates hear over 95 per cent of criminal environmental cases, a magistrate will hear an environmental

[63] For an overview of budget cuts affecting environmental regulators in Canada in the mid 1990s, see ch 4 s III. See ch 5 s III for a discussion of the impact this had on the number of prosecutions launched against environmental offenders.

[64] Meetings took place in October 2004. For a discussion of the real versus regulatory crime debate, see ch 6 s II C.

[65] Again, discussions took place in October 2004.

[66] Environmental Audit Committee Second Report of Session 2004–05, *Corporate Environmental Crime* (HC136) para 18.

[67] Macrory Review, Better Regulation Executive, *Regulatory Justice: Sanctioning in a post-Hampton World* (Consultation Document, May 2006) para 2.6.

[68] DEFRA, *Review of Enforcement in Environmental Regulation, Report of Conclusions* (October 2006), para 5.8.

case only on average once in seven years.[69] As judges will usually deal with mainstream crime such as theft and burglary, they 'find it difficult to know where to pitch environmental crime.'[70] Judges may not perceive environmental crime as real crime[71] and with most crimes dealt with by magistrates carrying a maximum £5,000 fine, they may baulk at using the higher maxima in environmental offences of £20,000.

There is, however, some indication that the courts in England and Wales are starting to treat environmental damage more seriously. In its 2006 Spotlight Report, the Environment Agency states that when dealing with serious, prolonged and deliberate criminal behaviour, the courts are responding and awarding fines which are more commensurate to the offence.[72] In 2006, successful prosecutions against companies led to over £3.6 million in fines, an average of £11,800 per business (compared with a total of £2.7 million and an average of £8,600 in 2005).[73] Furthermore, the recent DEFRA review also found evidence to suggest that the courts appear to take careful account of questions relating to seriousness, culpability and ability to pay, when sentencing environmental offenders.[74]

E Improving the Deterrent Impact of Fines: Practical Solutions and Equity Fines

The previous section outlined why fining an environmental offender in the criminal courts will not necessarily deter non-compliant behaviour. Under the Becker framework, the deterrence function of a fine is affected by its value (as discounted by the probability of prosecution and conviction) relative to the utility/benefit that the offender derives from committing the offence. From a practical perspective, fines are too low and prosecutions are rare. There are also theoretical limitations to fines, all of which restrict their deterrence function. But many of these limitations can be overcome by utilising alternative sentencing options which are better able to influence both individual and corporate behaviour. This is a matter to which I return in Section V below. Rather, in this section I review some of the ways in which law and policy-makers in Australia, Canada, and England and Wales have tried to influence the courts into awarding higher fines in pollution control cases and consider an alternative way of calculating fines that could improve their deterrent impact.

One option, which has been adopted by legislative authorities in Australia, Canada and, to some extent, England and Wales, is to raise the statutory maximums. Of course, merely increasing the maximum fine in the statute books will

[69] Information presented by the Environment Agency in oral evidence given to the Environmental Audit Committee.

[70] Above n 66 para 13.

[71] For a discussion of the real v regulatory crime debate see ch 6 s II C.

[72] Environment Agency, *Spotlight Report 2006*, 10.

[73] *Ibid.*

[74] Above n 68, para 5.9.

not necessarily lead to a corresponding increase in the fines imposed by the courts. And evidence on this matter is not conclusive either way. For example, a report carried out by WRc Plc for DEFRA (England and Wales) found 'no evidence that past hikes in the maximum lower court fine have led to any marked change in the general level of environmental fines'.[75] In contrast in its report on sentencing corporate offenders, the NSW Law Reform Commission found evidence that an increase in the statutory maximum applicable to the offence of water pollution was mirrored by the courts in the imposition of higher fines.[76] From a sample of 21 cases heard by the Land and Environment Court (LEC) NSW in 2000–01, the average fine for a water pollution offence under the former Clean Waters Act (NSW) was $15,912. In 1997, the PEOA 1997 (NSW) raised the maximum from $125,000 to $250,000 and 15 convictions of corporate offenders for the same offence under the new Act led to a higher average fine of $27,617. The Law Reform Commission concluded that 'the legislative increase of statutory maximum penalties to adjust the sentencing patterns of judges seems an effective method of achieving higher penalties.'[77]

The failure of the courts to make proper use of the full existing range of fines has also led many to call for the introduction of a specialist separate environmental court or tribunal.[78] Such calls are driven in part by a presumption that judges sitting in a specialist court like the LEC NSW (which has inter alia summary criminal jurisdiction) will impose commensurately higher fines than their counterparts in a local court. A statistical comparison between fines in local courts and the LEC would not be all that revealing. As indicated by the DECC (NSW) Annual Reports, the majority of cases heard by local courts concern matters such littering, smoky vehicles and defective vehicle noise control equipment. For these offences, fines are unsurprisingly relatively low and in 2005–06 were between $100 and $8,000.[79] In contrast, offences relating to, for example, water pollution, breach of licence conditions and the unlawful deposit of waste are heard by the LEC. Therefore, the higher fines imposed by the LEC arguably reflect the relative seriousness of these breaches, with fines ranging in 2005–06 from $500 to $25,000.[80] Interviews with regulatory officials in NSW also revealed a general sense of dissatisfaction with the sentencing practices of the LEC judges.[81] To date attempts to persuade the Victorian and UK governments to establish a specialist court have failed, possibly

[75] Above n 68 at Annex B, 9.

[76] Law Reform Commission (NSW) Report 102 (2003), *Sentencing: Corporate Offenders* ch 6 para 6.54.

[77] *Ibid* para 6.54.

[78] See, eg, in the UK M Grant, *Environmental Court Project: Final Report* (September 1999) available at <http://www.ingentaconnect.com/content/oup/envlaw/2001/00000013/00000003/art00423> accessed 5 January 2009, and M Woods and R Macrory, *Environmental Civil Penalties: A More Proportionate Response to Regulatory Breach* (University College London, 2003). Various models have been proposed but in all cases, the remit of the court would extend beyond the hearing of criminal prosecutions to include, eg, environmental appeals and judicial review.

[79] DECC, *Annual Report 2005–06*, 164–6.

[80] *Ibid.*

[81] Interview took place in March 2004.

due to the cost of establishing a court with specialist jurisdiction and the potentially very small case-load.

As discussed in Section C previously, one of the most important limitations to fines as applied to corporate entities is that they are small compared with the turnover of the company and are therefore seen merely as a 'cost' of doing business. One way round this is to fundamentally alter the way in which fines are calculated, such as through the use of equity fines. The name most closely associated with equity fines, Coffee, has described them as follows:

> When very severe fines need to be imposed on the corporation, they should be imposed not in cash, but in the securities of the corporation. The convicted corporation should be required to authorise and issue such number of shares to the state's crime victim compensation fund as would have an expected market value equal to the cash fine necessary to deter illegal activity. The fund should then be able to liquidate the securities in whatever manner maximises their return.[82]

In essence therefore, equity fines amount to a form of stock dilution. Instead of paying a monetary fine, the offending company is required by the court to issue a certain number of shares to an appropriate beneficiary, the value of which should amount approximately to the level of fine which would be necessary to deter the company from committing the offence. The beneficiary is then in a position to liquidate the shares in such a way as to maximise the return.

Equity fines arguably provide a more effective deterrent than traditional criminal fines. First, they avoid the deterrence trap. By requiring a company to issue shares, the 'wealth ceiling', which places an absolute limit on fines, is avoided. In other words, equity fines do not depend on corporate solvency. The sanction, which is not limited to current assets, also gives the beneficiary access to future earnings, ownership in plant and equipment and a stake in property investment. In sum, the deterrence function of equity fines is much greater than monetary fines. Second, any 'spillover' effect on consumers and employees is minimal and, rather, the brunt of the sanction is felt by shareholders. Indeed, where the senior management of the company holds a substantial number of shares (which is often the case) then equity fines may actually target the blameworthy actors in the firm. They could therefore have a direct deterrent effect on corporate management.[83] Remember though that equity fines may also affect 'innocent' shareholders who have no say in the running of the company.[84] This may make the courts more careful in sentencing and may, ironically, lead to lower penalties than if a monetary

[82] Above n 35, 413.

[83] This argument is of course weakened if this particular group of corporate personnel sell their shares in anticipation of the sanction. After all, they are in the best position to know the likely outcome of criminal prosecution.

[84] Kennedy argues that equity fines do not affect shareholders any more than monetary fines in that, in the long-term, a monetary fine may stifle growth or even bring the risk of bankruptcy. See C Kennedy, 'Criminal Sentences for Corporations: Alternative Fining Mechanisms' (1985) 75 *California Law Review* 443, 448.

fine were imposed.[85] A third, and related, point is that where a court issues shares to an appropriate third party, this brings new shareholders into the frame. The structure of the company may be affected, and this could increase the likelihood of compliance. Fourth, and in contrast with a cash fine, an equity fine is more difficult to write off as a mere cost of doing business.

It is also useful to think about the extent to which equity fines affect the non-financial values of a company, ensure individual accountability or direct corporate reform. In relation to the former, it is arguable that equity fines could affect corporate reputation and power even though they are designed primarily to have a punitive financial impact. They are also more likely to ensure individual accountability within the firm as shareholders could initiate internal disciplinary action against the guilty parties although there is of course no guarantee that such action would be taken. Finally equity fines, because of their greater deterrent impact, are more likely than monetary fines to generate corporate reform of operating and monitoring procedures although again, this is not guaranteed. As summarised by Jefferson:

> The amount of the equity fine may be sufficient to stigmatise the company but would have to be very large to lead to a takeover: there is no guarantee that shareholders will exert pressure on managers; and there is also no guarantee that corporate malpractices will be put right.[86]

Equity fines also have one other important advantage. By requiring the shares to be allocated to a suitable beneficiary, appointed by the court, the sanction could potentially compensate victims of the offence. There is, of course, nothing to stop monetary fines being used in this way but an equity fine is more likely to go to victims.

Saying all that, there are a large number of problems in using equity fines as a criminal sentencing tool. One of their biggest drawbacks is that the imposition of an equity fine may be unsuitable in many cases, especially those involving small, privately held companies that involve closely held securities in family-type arrangements.[87] It is likely that a large proportion of pollution offences are committed by such entities. So although equity fines do have some major advantages over monetary fines, particularly in terms of their deterrent effect, they are not a panacea. Of course, if used in conjunction with other sentencing mechanisms such as probation orders, then many of these drawbacks could potentially be countered.

F Conclusion

In Australia, Canada, and England and Wales, fines are the sanction used most frequently by the criminal courts when sentencing environmental offenders.

[85] Saying that, investing in a share portfolio carries with it a number of risks. If the company benefits financially from a criminal offence, then why should any shareholder profit from that?

[86] M Jefferson, 'Corporate Criminal Liability: The Problem of Sanctions' (2001) 65 *Journal of Criminal Law* 235, 248.

[87] Above n 76, ch 7 para 7.26.

However, from a practical perspective, their deterrent impact is at best questionable. Using Becker's framework, if the low level of fines is combined with the infrequency of prosecution, then the expected penalty (the sanction discounted by the likelihood of prosecution) is unlikely to outweigh the benefits of non-compliance. Even if fines were higher and prosecutions more frequent, their deterrent impact remains limited. First there is the 'deterrence' trap; that is that the maximum fine imposed is constrained by the wealth of the offender, a particular difficulty in the case of corporate entities for whom many alternative sentencing options do not apply. And second, by virtue of the fact that the financial consequences of a fine may be borne by third parties such as shareholders and employees, they may not have an impact on responsible officers in the corporation. Fines may also not be effective in delivering other sentencing goals such as rehabilitation and retribution: they do not compel a company to change its behaviour and may not be seen to deliver justice.

All in all, fines alone are not capable of delivering the many goals of the criminal justice system, including deterrence. The courts in Australia, Canada, and England and Wales can turn to alternative traditional criminal penalties such as incarceration and community-based penalties. These matters are the focus of Section IV below. In Sections V and VII consider the creative sentencing powers of the courts in Australia and Canada when dealing with polluters. These creative options can be used in conjunction with more traditional penalties to strengthen the deterrence signal sent out by the criminal justice system.

IV Restricting the Freedom of Individual Offenders

This section examines the availability and use of criminal sanctions which protect the public by restricting the freedom of individual polluting offenders. Unlike fines, they impose a non-financial disutility on the offender and as a result their deterrent impact, as assessed by the economic framework developed in Chapter 2, is more difficult to evaluate. Although it is generally accepted that custodial sentences send a very strong deterrence signal, other sanctions falling within this category such as community penalties may not. This could go some way to explaining why the courts seem reluctant to use these sanctions when sentencing polluters. Probably one of the most important sanctions in pollution control regulation that is aimed at incapacitating offenders is licence revocation and suspension. The power to revoke or suspend a licence is normally administrative as opposed to judicial, and so is discussed in Chapter 8.

A Custodial and Suspended Sentences

Custodial sentences are commonly available to the courts in sentencing environmental offenders but their use is relatively unusual and would seem to be reserved for only the most serious of cases. Both provincial and federal Canadian environmental law statutes make provision for the imposition of custodial sentences.[88] Whilst this sentencing power has been utilised by the Canadian courts, examples are rare, indicating that jail terms are usually inappropriate where an environmental offence has been committed. In Ontario, for example, from November 2001 to November 2003, custodial sentences totalled nine months. A search of the Ministry's news releases and court bulletins reveals that a custodial sentence was last imposed in September 2003. Daher Wanna, the owner of 31 Auto Parts, a car wrecking company, was sentenced to three months in jail for illegally discharging pollutants, some of which contaminated surface water and soils at the company's site in Osgoode, Toronto.[89] The company was also found guilty of illegal waste practices and failing to comply with a provincial officer's order.[90] It would appear that jail sentences are used most commonly in prosecutions involving wildlife and fisheries legislation. For example in several instances, Environment Canada has secured a custodial sentence in cases involving the illegal smuggling of wild birds such as tropical finches and eagles into Canada.[91]

Under Australian pollution control regulation, courts also have the power to imprison individual offenders[92] although the courts must consider (and reject) all possible alternatives before considering imprisonment as appropriate.[93] Custodial sentences may be suitable where for example, the offender's conduct involves a considerable degree of wilfulness or deception, is over a long period of time or repetitive in nature, or where there is a high risk of harm to the public or the environment.[94] Despite the statutory availability of custodial sentences, from 2003–06 no polluters in Victoria and NSW faced a custodial sentence.[95] This could of course be explained by the fact that the majority of cases do not cross the threshold of criminality necessary to justify a custodial sentence. In NSW for example, only three prosecutions since 2003 have related to Tier 1 offences.

[88] See, eg, CEPA (federal) s 272(2) and EPA 1990 (Ontario) s 187(1).

[89] The company and the owners of the site also received substantial fines.

[90] See <http://www.ene.gov.on.ca/envision/news/2003/100301.htm> accessed 5 January 2009. A provincial officer's order is an example of an administrative enforcement notice. They are discussed further in ch 8 s III A iii.

[91] See, eg, the prosecution of three members of the Flikkema family who imported tropical birds from Africa, through Europe, into Canada and the US. Environment Canada Press Release July 7, 2000.

[92] See, eg, PEOA 1997 (NSW) s 119 and EPA 1970 (Victoria) s 59.

[93] Crimes (Sentencing Procedure) Act 1999 s 5(1), *R v James* (1985) 14 A Crim R 364, and *R v O'Connor* (1986) 23 A Crim R 50, 54–5. In Victoria, see Sentencing Act 1991 s 5(4), which states that 'A court must not impose a sentence that involves the confinement of the offender unless it considers that the purpose or purposes for which the sentence is imposed cannot be achieved by a sentence that does not involve the confinement of the offender.'

[94] Above n 18, 157–8.

[95] According to Annual Reports. The first custodial sentence for an environmental offence in Australia was imposed by the Court of Petty Sessions in Western Australia in 1995. Mr McMurty, a company director, was sentenced to three months' imprisonment for knowingly allowing 22,000 litres of toxic chemicals to be dumped into a drain, causing the death of plants, animals and fish. See *EPA v McMurty* unreported, Court of Petty Sessions, WA, 9 March 1995. Cited in Hain and Cocklin above n 52, 332.

Nearly all pollution control offences in England and Wales are also punishable by imprisonment, with maximum prison terms prescribed in the legislation (normally three months on summary conviction and two years on indictment).[96] Compared with their counterparts in Australia and Canada, it would seem that the English courts use custodial sentences more frequently. In 2006, the courts sentenced 10 individual defendants to a total of 138 months in prison, the majority of which related to waste offences. This figure is significantly higher than in previous years although the following year, in 2007, only six individuals were given a custodial sentence, amounting to a total of 93 months' imprisonment.[97]

In some circumstances, the court may also impose a suspended sentence whereby the judge suspends the passing of a sentence for a fixed period. If that period expires without further offences, there will be no sentence.[98] The extent to which the courts use these powers would appear to vary from one jurisdiction to the next. The annual reports of environmental regulators in NSW and Victoria provide no examples of suspended sentences. The Ministry of Environment (Ontario) does not produce any annual statistics relating to enforcement, although it is clear from reading the publicly available Court Bulletins that suspended sentences are used for substantive pollution offences and would appear to be more frequent than custodial sentences. In contrast, the courts in England and Wales imposed only six suspended sentences from 2004–06, considerably less than the 26 defendants sentenced to imprisonment.[99] However in 2007, the number increased and 11 defendants have had their custodial sentences suspended. This included one company director.[100]

There are a number of benefits to incarceration namely, deterrence, rehabilitation, incapacitation and retribution. From a deterrence perspective, the seriousness of the sanction may deter the offender and others from committing future crimes, although as imprisonment is used so rarely in pollution cases, its deterrent impact (as an expected penalty) will be extremely limited. There is also some doubt as to the deterrent effect of imprisonment more generally, leading commentators such as Ashworth to conclude that 'the preventive effects of custody are frequently overestimated.'[101] But Brian Preston the Chief Judge in the NSW LEC is supportive of the deterrent impact of custodial sentences, particularly when applied to corporate officers for whom 'the risk of imprisonment for criminal liability has a deterrent effect and provides a greater incentive to ensure corporate compliance.'[102]

[96] See, eg, Water Resources Act 1991 s 85. Custodial sentences for waste offences have recently been extended to 12 months (magistrates' court) and five years (Crown Court). See Clean Neighbourhoods and Environment Act 2005 s 41.

[97] Information taken from Environment Agency *Spotlight Reports*. Details provided in Table VIII below.

[98] See, eg, Provincial Offences Act 1990 (Ontario) s 72(1), Crimes (Sentencing Procedure) Act 1999 (NSW) s 12, and Powers of Criminal Courts (Sentencing) Act 2000 (England and Wales) s 118.

[99] Information from Spotlight Reports.

[100] Environment Agency, *Spotlight on Business: 10 Years of Improving the Environment* (Environment Agency, July 2008) 21.

[101] A Ashworth, *Sentencing and Criminal Justice*, 3rd edn (London, Butterworth, 2000) 240.

[102] Above n 18, 156.

A second possible rationale for imprisonment is rehabilitation although in some countries such as the USA and the UK, the idea of rehabilitation has fallen out of favour.[103] In a very general sense, rehabilitation can amount to deterrence in that an offender, on being imprisoned, does not commit crimes on release. More specifically, rehabilitation may refer to the opportunities an offender has in prison to learn new social values and skills. The reformative potential of prisons is, however, debatable and no criminological study has yet been designed to test whether imprisonment increases an offender's criminal know-how. In the UK in the early 1990s, Lord Woolf's inquiry into prison disturbances concluded that:

> It is now generally accepted that, particularly with young offenders, there is a risk that a custodial sentence, instead of making it less likely that the offender will offend again, increases that danger.[104]

However, it has been argued that the side-effects of prison are prone to exaggeration. Imprisonment may also provide a form of retribution for society, in that where people knowingly and wrongly injure other persons or interests, then justice is served through punishment. It is, however, very difficult to even estimate the 'value' of this retributive element.

Probably the most obvious benefit of imprisonment is incapacitation. During an offender's time in prison, he is unable to re-offend. From an economics perspective, this provides a net benefit as due to high enforcement and punishment costs, preventing a crime by incapacitating the potential offender saves the state the money associated with both catching and punishing the offender, and the net damage caused by the crime (which is almost always positive). However, the high costs of incapacitation, especially compared with other non-custodial sentencing options, need to be justified by real reductions in crime rates. Easton and Piper comment that:

> There is also the question of whether prison does incapacitate effectively if it ultimately leads to higher offending rates than non-custodial sentences because of factors such as the stigmatising effects of prison, the difficulty in obtaining employment on release, the effects of prison dehumanisation on the offender, and the opportunities to learn from other offenders . . . even if we could establish that imprisoning more offenders and increasing sentence length leads to a reduction in crime, we would need to assess the economic burdens.[105]

Available research suggests that in order to achieve even a small reduction in crime rates, there would have to be a substantial increase in the use of custody.[106] Incapacitation may not therefore be cost effective. However, as a crime control strategy, incapacitation has one clear advantage over deterrence and rehabilitation.

[103] For a brief summary, see Ashworth above n 101, 239.
[104] Lord Justice Woolf, *Prison Disturbances, April 1990: report of an Inquiry* (HMSO) para 10.25.
[105] Above n 7, 133.
[106] For a summary of some of the available literature, see Easton and Piper *Ibid* 133.

Deterrence only works if potential offenders take into account the costs and benefits of their unlawful action. Rehabilitation will only be successful if the social values of the offender can be altered. In contrast, incapacitation works by definition.

B Community Penalties and other Non-Custodial Sentencing Alternatives

Sentencing laws in Australia, Canada, and England and Wales provide for a variety of community-based and other penalties as alternatives to imprisonment.[107] Unlike most of the creative sentencing options discussed in Sections V and VI, the penalties discussed in this section can normally only be used in sentencing individuals.

Polluters may be subjected to community service. In England and Wales, for example, offences committed prior to 4 April 2005 were dealt with by Community Punishment (Service) Orders (which required the offender to do up to 240 hours unpaid, supervised work on local community projects), Community Rehabilitation Orders (a community sentence which involved regular contact with the Probation Service) and a combined Community Punishment and Rehabilitation Order.[108]

As can be seen from Table VIII below, from 2004–07, Community Punishment (Service) Orders were used far more frequently than custodial sentences, indicating the value the courts place on the carrying out of community service in environmental cases.[109] In a small number of cases, company directors were required

Table VIII: Custodial Sentences and Community Orders (Number of defendants)[110]

	Custodial Sentence	Community Punishment Order	Community Rehabilitation Order	Community Punishment and Rehabilitation Order
2004	12 (36 months)	42 (284 days)	2 (18 months)	2 (24 months)
2005	4 (18 months)	25 (152.5 days)	1 (12 months)	1 (2 years)
2006	10 (138 months)	23 (243.5 days)	0	0
2007	6 (93 months)	19 (172 days)	0	1 (100 hours)

* total number of hours/days/months/years in brackets

[107] For example, community service and punishment, probation and good behaviour bonds and conditional discharges.

[108] See s 46, 41 and 51 of the Powers of Criminal Courts (Sentencing) Act 2000. For an insight into the development of community service in the UK, see K Pease, 'Community Service Orders' in N Morris and M Tonry (eds), *Crime and Justice: An Annual Review—Vol 6* (Chicago, University of Chicago Press, 1988) and AA Vass, *Alternatives to Prison: Punishment, Custody and Community* (London, Sage Publishing, 1990) ch 5.

[109] These orders are not the same as environmental project orders, discussed in s VI D, which direct that the work undertaken relates to the offence or to the environment more generally and can be used against corporate entities as well as individuals.

[110] Information taken from Environment Agency Spotlight Reports 2004–07.

to comply with a community penalty. For crimes committed after 4 April 2005, section 177 of the Criminal Justice Act 2003 introduced a single community order that can be tailored to reflect the sentencing requirements of individual offenders. So, for example, a community order can require a convicted person to carry out unpaid work, an activity or a programme. Such an order was made in a prosecution brought by the Agency in July 2008 against the owner of a company who pleaded guilty to two offences of allowing his van to be used to fly-tip waste. He was sentenced to a Community Order, requiring him to complete 150 hours of unpaid work for the community for each offence, and was ordered to pay a contribution of £250 towards the costs of bringing the case.[111] Community service is also available in Australia[112] although it would seem that it is rarely employed in sentencing for environmental offences.[113]

There are a range of other alternatives to imprisonment. Probation, which will normally require an offender to meet up with a probation officer at regular intervals, is usually available to the courts when sentencing polluting offenders.[114] In Canada, probation orders, which can be used in conjunction with a community service requirement, are more common than custodial or suspended sentences.[115] In contrast, probation is rare in environmental cases in Australia and England and Wales. Other sentencing options include absolute or conditional discharges. These allow a court to find the defendant guilty but, where punishment is not appropriate, the offender can be discharged absolutely or conditionally (on the condition that the offender does not commit a crime during a set period of time).[116] Unlike community service and probation orders, conditional discharges can normally be used in the case of companies. Data produced by the Environment Agency (England and Wales) indicates that, in 2006, 45 defendants (including nine companies and two company directors) were conditionally discharged for a total of 1,492 months.[117] In 2007, a total of 57 defendants were conditionally discharged; 47 individuals, eight companies and two company directors.[118]

[111] See Press release available at: <http://www.environment-agency.gov.uk/news/86216.aspx?page=11&month=9&year=2008> accessed 7 January 2009.

[112] See, eg, Crimes (Sentencing Procedure) Act 1999 (NSW) s 8.

[113] Above n 18, 158.

[114] See, eg, Provincial Offences Act 1990 (Ontario) s 72(2), (3). In NSW, a probation requirement can be attached to a good behaviour bond. See Crimes (Sentencing Procedure) Act 1999 s 9. Good behaviour bonds can also be entered into as part of a conditional discharge (Crimes (Sentencing Procedure) Act 1999 s 10) and may be applied to companies, even though the relevant provisions are 'referable only to non-corporate persons'. (see *Environment Protection Authority v Virotec International Ltd* [2002] NSWLEC 110, [36]). In England/Wales see Powers of Criminal Courts (Sentencing) Act 2000 s 41.

[115] See Ministry of Environment Press Releases. There is some evidence that probation orders can be used against corporate entities, despite the legislative provisions indicating otherwise. See further *R v Bata Industries Ltd* (1995), 25 OR (3d) 321.

[116] See, eg, Powers of Criminal Courts (Sentencing) Act 2000 (England and Wales) s 12, and Crimes (Sentencing Procedure) Act 1999 (NSW) s 10.

[117] Environment Agency, *Spotlight on Business: Environmental Performance in 2006* (July 2007) 10.

[118] Above n 100.

V 'Creative' Sentencing Powers—An Overview

A Introduction

So far, I have examined what could be described as the 'traditional' criminal sentencing tools. I have questioned the extent to which fines, imprisonment and non-custodial alternatives to imprisonment can deter both individuals and, where relevant, companies, taking into account the costs that these sanctions impose on society. Fines, which represent the starting point in sentencing environmental offenders, are limited in their ability to deter offenders, especially where the offender is a corporate entity. Although their deterrent impact could be strengthened by, for example, calculating them as a percentage of turnover, profitability or even liquidity, creative accounting can be used to mask the true financial position of business. Furthermore, fines are not effective in internalising pollution costs. Not only is it difficult to calculate the financial penalty that will internalise costs by making the polluter pay, but doing so only on those occasions that the polluter is caught will not take into account all those times when he is not. So, from an economics perspective, fines alone are not capable of deterring offenders and promoting compliance. Alternative sentencing options may better address non-compliance and may send stronger deterrence signals.

A second rationale that supports creative sentencing in corporate criminal liability in particular relates to the non-financial values in organisational decision making.[119] It is generally accepted that the cost-benefit calculation on which our deterrence framework is based is more significant in corporate as opposed to individual criminal law. Corporate acts are guided by the need to reap some kind of economic benefit and fines, as the primary deterrent sanction, most obviously address any financial reward accruing from non-compliance. But, according to Fisse, corporations have non-monetary as well as monetary motivations. The individual desire for money, power and prestige can all drive a company in a particular direction, and whilst 'in practice, pursuit of any one tends to promote the other'[120] deterrent-based 'creative' sanctions such as corporate probation and formal adverse publicity requirements may be deployed to directly affect those non-financial values. Fisse also argues that rehabilitation, incapacitation and retribution are, in the context of corporate criminal law, subgoals of deterrence. When corporate entities are punished or threatened with punishment, the message conveyed is catalytic as well as inhibitory, in the sense that companies are being directed to guard against committing the same offence again. This, Fisse argues, is peculiar to corporate offences:

[119] See generally Fisse above n 4, 1145–59.
[120] *Ibid* 1155.

> The principal reason for this difference is that organizational offenders cannot exert self-control merely by individual self-denial. Self-denial on these offenders' parts must be embodied in corporate policy and backed by appropriate disciplinary measures and organizational procedures . . . punishment or threat of punishment requires corporations to do more than merely exercise inhibition and self-restraint; they are expected to institute effective crime prevention policies, disciplinary control and changes in standard operating procedures.[121]

The organisational responses Fisse refers to are all forms of rehabilitation and partial incapacitation. In other words, prosecution (or the threat of it) can encourage a company to introduce measures which will reduce the opportunity for corporate crime.

A third and final argument in favour of creative sentencing is what Scott Campbell refers to as the corporate reformist movement.[122] This movement advocates the view that 'something more than fines are necessary in environmental sentencing, and that the economics approach will never by itself achieve a compliance culture.'[123] Non-monetary sanctions will be more effective in preventing both individuals and companies from passing on environmental violations as a mere cost of doing business. The ethical underpinnings of the corporate reformist movement are also very different from the economics approach. In the case of the former, environmental sentencing is designed to reflect aesthetic, non-anthropocentric justifications for environmental protection, as opposed to anthropocentric, utilitarian ones. Fines and other traditional criminal penalties can only apply to humans and their organisations, whereas non-fine measures such as environmental project orders, can address the environment more directly and so can reflect the non-economic, non-human character of environmental damage. In sum, either directly or indirectly, all three rationales can promote creative sentencing options on the basis that they improve and strengthen the deterrent impact of the criminal law.

B The Development of Creative Sentencing

In Australia, the low level of criminal fines, coupled with their potentially limited deterrent effect, prompted state legislators including those in NSW and Victoria to provide the courts with a more diverse range of criminal penalties that could be tailor-made to the specifics of the offence and the offender.[124] As aptly put by the NSW Minister of the Environment during the second reading of the Protection of

[121] See generally Fisse above n 4, 1160.

[122] Above n 32, 6–7.

[123] *Ibid.*

[124] Some of the alternative sentencing options are not available to all courts. For example, many sentencing options under the PEOA 1997 (NSW) can only be awarded by the LEC. At federal level, the compliance and enforcement provisions of the Environmental Protection and Biodiversity Conservation Act 1999 were strengthened by the Environment and Heritage Legislation Amendment Act (No 1) 2006. However, the majority of changes relate to administrative as opposed to judicial and enforcement powers and, as a result, the Act will not be discussed in this chapter.

the Environment Operations Bill which introduced a number of alternative criminal penalties:

> We are working to broaden the options available to the courts. We want changed behaviour and improved environmental performance and are giving the courts an opportunity to teach a salutary lesson to those who have been found guilty.[125]

Although there is still provision for the imposition of monetary fines, both alternative monetary and non-monetary sanctions such as environmental service orders and adverse publicity orders are becoming more prevalent and there is evidence to suggest that the courts are starting to embrace the concept of alternative sentencing, particularly in its application to corporations. In all cases, the court orders available can be imposed in addition to a fine, and one or more orders can be made against an offender. The most advanced state in this respect is NSW which makes provision for a wide range of alternative sentencing options.[126] Non-compliance with a creative sentencing order awarded by the NSW LEC is a criminal offence subject to a maximum daily fine of $120,000 for corporations and $60,000 for individuals.[127] Under the Victorian EPA 1970, failure to comply with an order is contempt of court and the regulator:

> (a) may do anything that is necessary or expedient to carry out any action that remains to be done under the order and that it is still practicable to do; and
> (b) may publicise the failure of the person to comply with the order.[128]

The tables below illustrate the use of alternative sentencing options, relative to fines, in NSW and Victoria.[129]

In Canada, most federal and provincial environmental statutes extend the remedial powers of the criminal court beyond fines and imprisonment. Initial calls for the introduction of additional sentencing measures were made by the judiciary, most notably Stuart CJ in the case of *R v United Keno Hill Mines Ltd*, who considered that:

> A greater spectrum of sentencing options is required to ensure effective deterrence and prevent illegal economic advantages accruing to corporations willing to risk apprehension and swallow harsh fines as operating costs.[130]

The range of creative sentencing options in Canada is broad. The EPA 1990 (Ontario) explicitly empowers the courts to award, for example, monetary benefits, restoration and restitution orders, non-compliance with which is a criminal

[125] Second Reading of PEO Bill 13 November 1997, 1836.
[126] In NSW, a local court is not authorised to make certain orders (PEOA 1997 s 250(1)).
[127] PEOA 1997 (NSW) s 251.
[128] EPA 1970 s 67A(5).
[129] Creative sentencing powers available in the South Australian Environment Protection Act 1993 have yet to be used by the courts. See further, D Cole, 'Creative Sentencing—Using the sentencing provisions of the South Australian Environment Protection Act to greater community benefit' (2008) 25 *Environmental and Planning Law Journal* 13.
[130] (1980) 10 CELR 43, para 38.

Table IX: Cases before the Land and Environment Court (NSW)—Prosecutions completed under EPA Legislation[131]

	2003–04	2004–05	2005–06	2006–07
Fines	39	18	17	16
Adverse Publicity Order	14	6	6	5
Restoration and Prevention Order	4	5	0	0
Costs relating to restoration and prevention	1	0	0	0
Project Order	0	1	2	2
Environmental Audit Order	0	0	0	1
Compensation	0	1	0	0
Monetary Benefits Order	0	0	0	0
Investigation Costs *	1	3	7	2
Other	2	0	0	1
Total Number of Prosecutions	45	28	20	19

* The Annual Reports do not specify whether legal costs are awarded to the prosecution. However, a perusal through the LEC judgements would indicate that in the overwhelming majority of cases, such costs are awarded.

Table X: Results of Major Prosecutions completed under EPA Legislation (Victoria)[132]

	2003–04	2004–05	2005–06	2006–07
Fines	3	8	4	3
Adverse Publicity Order	4	3	4	8
Restoration and Prevention Order	0	0	0	0
Costs relating to restoration and prevention	0	0	1	1
Project Order	5	5	5	9
Environmental Audit Order	0	0	0	0
Compensation	0	1	0	0
Costs*	9	7	5	11
Good Behaviour Bond	2	1	4	
Other	1	0	0	1
Total Number of Prosecutions	9	9	12	13

* The Annual Reports do not distinguish between legal and other costs

131 Information taken from EPA/DECC Annual Reports.
132 Information taken from EPA (Victoria) Annual Reports.

offence.[133] Section 190 of the EPA 1990 also gives the courts a general power to attach other conditions which relate:

> [T]o the circumstances of the offence and of the person that contributed to the commission of the offence as the court considers appropriate to prevent similar unlawful conduct or to contribute to rehabilitation.

This provision gives the courts broad discretion in tailoring their sentencing to the individual facts and circumstances of the case and its operation is neatly illustrated by recent decisions of the Ontario courts.[134] In 2004 for example, the company officer of Achievor Recovery Ltd was convicted of breaching the conditions of a site certificate of approval relating to the processing and storage of white goods and used tyres. The officer was fined $25,000 and a section 190 order was made prohibiting him from engaging in the tyre recycling business anywhere in Ontario.[135] Also in 2004, an order was made under an equivalent provision of the Ontario Water Resources Act against Chevalier Engineering Associates Ltd. The company pleaded guilty to two counts of failing to notify the Ministry of Environment and the Medical Officer of Heath about the turbidity and bacteriological exceedances in treated water analysis results. In addition to a $20,000 fine, it was ordered, under section 112, to cease advising or providing consulting or professional services to any waterworks in the provinces.[136]

There are a host of other orders provided for in both Canadian federal and provincial environmental legislation. Specific provision is made at federal level in CEPA and the Fisheries Act for the making of, inter alia, environmental management orders, environmental audit orders, orders requiring the funding of trust funds, scholarships and environmental programmes and orders requiring investment in environmental research.[137] Again, non-compliance is a criminal offence, and will be subject to the same penalties as would apply to substantive pollution offences.[138]

At the time of writing, the courts in England and Wales have, compared with their counterparts in Australia and Canada, much more limited sentencing powers. Whilst legislation provides for inter alia fines, custodial sentences, community penalties, costs orders and, in some cases, remediation orders, the courts do not have at their disposal the wealth of creative sentencing options available in other jurisdictions. However, it would seem that the time is ripe for reform. Environmental law enforcement has been the subject of two recent reviews carried out by Professor Richard Macrory (on behalf of what is now the Department for Business Enterprise and Regulatory Reform) and the Department of the

[133] See EPA 1990 ss 189 (monetary benefit order), 190 (order to prevent damage) and 186(2).

[134] Examples of how this section of the EPA 1990 has been used are provided in the following sections.

[135] Information provided by Ministry of Environment, Ontario.

[136] Information provided by Ministry of Environment, Ontario.

[137] See, eg, CEPA (Canada) s 291 .

[138] See CEPA (Canada) s 272(1)(d).

Environment, Food and Rural Affairs.[139] The consultation and review processes led to calls for the introduction of wider, creative sentencing powers such as profit orders, corporate rehabilitation orders and community project orders. Evidence received as part of the Macrory Review indicated that:

> Whilst recognising that more sentencing options exist but are not currently being used on a widespread basis—the current range of sanctioning options available for regulatory offences is too narrow.[140]

The House of Commons Environmental Audit Committee has also considered the question of environmental crime, with one report concluding that 'the criminal justice system currently offers no effective deterrent.'[141] There is no doubt that in considering the introduction of alternative sentencing options, the government can learn much from the experience of creative sentencing in other countries.

V Creative Sentencing Powers—Theory and Practice

A Introduction

In this section, I consider both monetary and non-monetary creative sentencing powers. Whilst all criminal sanctions have a deterrence-based function, and can therefore influence the deterrence framework developed in Chapter 2, they also reflect alterative sentencing rationales (such as retribution, restoration and rehabilitation) which are valuable, particularly in the context of corporate crime.[142] I begin in Section B by considering sanctions which are clearly financial in nature. Section C examines publicity-based sanctions such as adverse publicity orders. Much debate has surrounded the extent to which adverse publicity tackles both the financial and non-financial motivations of firms. In Section D, I examine the notion of corporate community service, more specifically the use of environmental service or project orders. And Section E concludes by evaluating rehabilitative sanctions such as environmental audit orders. Each section considers these sanctions from a theoretical perspective, examines legislative provisions in Australia, Canada and (where relevant) England and Wales, and provides, where possible, an insight into the courts' use of creative sentencing in pollution cases.

[139] R Macrory, *Regulatory Justice: Sanctioning in a Post-Hampton World—Consultation Document* (May 2006), R Macrory, *Regulatory Justice: Making Sanctions Effective—Final Report* (November 2006). Referred to hereafter as the Macrory Review. DEFRA, *Review of Enforcement in Environmental Regulation: Report of Conclusions* (DEFRA, October 2006).

[140] *Ibid* (Macrory Consultation) para 6.13.

[141] House of Commons Environmental Audit Committee, *Corporate Environmental Crime* HC (2nd Report of 2004–05) 136 [37]. See also House of Commons Environmental Audit Committee, *Government Response to the Committee's Second Report of Session 2004–05 on Corporate Environmental Crime* HC 434, and House of Commons Environmental Audit Committee, *Environmental Crime and the Courts* HC (6th Report of Session 2003–04) 126.

[142] See generally s II above.

B Monetary Sanctions

The monetary sanctions discussed in this section perform an important deterrence function. They can be awarded in addition to a fine and the costs of complying with the order must be borne by the offender. As a result, their value should be added to the value of D (the court-based sanction) in the deterrence framework developed in Chapter 2. However, the deterrence value of these orders is lessened by the fact that they are rarely used in practice. Furthermore, as they impose a financial disutility on the actor, they suffer from many of the weaknesses of fines.[143] They do not avoid the deterrence trap, they do not necessarily lead to operational changes and improved performance and can be seen as merely a cost of doing business. Their value may however lie in the fact that they reflect a number of different sentencing goals including reparation, restoration and retribution. By ensuring inter alia that regulators and victims are compensated for their losses, they increase social welfare and, from this perspective at least, are valuable in economic terms.

i Costs and Compensation Orders

Australian, Canadian and English law makes provision for the payment of the prosecutors' legal costs by convicted offenders.[144] In NSW, the PEOA 1997 also enables the LEC to award an investigation costs order and in most cases, the regulator will seek to recover these expenses.[145] However, it would seem that the LEC is reluctant to use these orders, with the DECC recouping investigation costs in only two out of the 19 prosecutions in 2006–07.[146] In Victoria, section 66C(2) of the EPA 1970 states that in addition to awarding legal costs to the Authority, the court can require the committed defendant to pay the 'reasonable market cost of any work conducted by the Authority' including costs incurred in testing, evaluating and recording evidence. According to EPA (Victoria) annual reports, costs are awarded in the majority of major prosecutions. The Environment Agency (England and Wales) will also seek to recover the costs incurred in investigating an offence in all cases and the majority of prosecutions brought by the Agency result in an investigation costs order.[147]

In addition, courts can order environmental offenders to pay compensation to either the regulator or an affected third party. A regulator or third party may seek such an order where it has suffered personal injury, loss or damage to property or,

[143] See s III C.

[144] See, eg, Criminal Procedure Act 1991 (NSW) s 253(2) and Magistrates' Court Act 1989 (Victoria) s 131, Provincial Offences Act 1990 (Ontario) s 60 and Prosecution of Offences Act 1985 (England and Wales) s 18.

[145] PEOA 1997 (NSW) s 248.

[146] See Table IX above.

[147] Environment Agency, *Enforcement and Prosecution Policy* (November 1998) para 27. See Environment Agency News Releases egs of costs orders.

in some cases, has incurred costs and expenses in preventing or mitigating (or attempting to prevent or mitigate) such loss or damage.[148] Such measures therefore come close to integrating criminal and civil remedies. In Ontario, Canada, compensation is secured through a 'restitution order'.[149] So, for example, the owner of a rug dyeing and cleaning company was prosecuted by the EPA (Ontario) in 2004 for permitting the fly-tipping of rusted and leaking drums of dyes and other chemicals at five separate locations in the Melancthon Township. In addition to a suspended sentence, a restitution order was issued, requiring the defendant to pay the Melancthon Township $8,000 for the costs incurred in cleaning up the waste.[150] As illustrated by the case of *R v Thompson Agricultural Aviation Ltd*, an important limitation to restitution provisions such as those discussed above is that the Canadian courts are reluctant to use them where the quantum of costs is disputed.[151] The case arose when the defendant company carried out aerial pesticide spraying in such a way as to damage trees located on the property of two local residents. The company was fined $2,000 but the court refused to award a restitution order. Before the case came to trial, one of the property owners sold her house and the court found that there was insufficient evidence which would enable the court to make a meaningful calculation of real loss. In such circumstances, the court will generally encourage the parties to proceed with a civil action.

As with many of the orders discussed in this section, compensation orders are rarely awarded by the courts. For example, from 2004–07 only six compensation orders were secured against environmental offenders in England and Wales although most recently in 2007, a total of four compensation orders were awarded.[152] From 2003–07 a total of two compensation orders were secured for breach of EPA legislation in both NSW and Victoria.[153] One recent example from the LEC in NSW is the case of *EPA v Obaid* in which the defendant was successfully prosecuted for unlawfully using land as a waste facility.[154] The defendant was fined $39,000 and ordered to pay $17,371.20 in compensation to the landowner for expenses incurred in removing the waste. Where the offence relates to the failure of the defendant to pay fees or charges to a regulatory authority, compensation orders can also be used by regulators to recoup this lost income. For example, in November 2008, the Environment Agency (England and Wales) received £9,122 in compensa-

[148] See, eg, Sentencing Act 1991 (Victoria) s 86(1) with EPA 1970 (Victoria) s 65A, and PEOA 1997 (NSW) ss 246 and 247. In England and Wales, see Powers of Criminal Courts (Sentencing) Act 2000 s 130, under which compensation can be awarded for any personal injury, loss or damage resulting from the offence.

[149] See, eg, Ontario Water Resources Act s 112.1, which allows the court to order restitution 'against the person convicted of the offence, requiring the person to pay another person for the reasonable expenses actually incurred by the other person on account of damage to property in which the other person has an interest that results from or is in any way connected to the commission of the offence, in such amount and on such terms and conditions as the courts considers just.'

[150] Information provided by EPA officer during interviews in October 2004.

[151] (2002) 47 CELR (NS) 155 (Alberta Provincial Court).

[152] See Environment Agency *Spotlight on Business* reports.

[153] See Tables IX and X above.

[154] [2005] NSWLEC 171.

tion for loss of income that it would have received if the defendant company had registered correctly under the Packaging Regulations.[155] By using compensation orders in this way, the criminal courts in England and Wales can go some way to removing any profit accruing to the offender from the offence. This would arguably provide an alternative, in some cases, to profit orders discussed below.

ii Restoration, Remediation and Prevention Orders

Many courts can order a convicted offender to prevent, mitigate or remediate any environmental harm that has occurred or is likely to occur, and to prevent the offence from continuing or recurring.[156] As with the 'costs' orders discussed above, restoration, remediation and prevention orders reflect reparative and restorative goals, as well as deterrence. In NSW, for example, restoration and prevention orders under section 245 of the PEOA 1997 (NSW) can require the offender to take such steps to prevent, control, abate or mitigate any harm to the environment caused by the commission of the offence and to make good any resulting environmental damage. Such an order may, for example, require the offender to remove and dispose of waste at a licensed facility, and furnish the regulator with proof of the lawful disposal.[157] The LEC may also require the offender to take such steps as to prevent the continuance or recurrence of the offence.[158] This is neatly illustrated by the recent case of *EPA v Ramsey Food Processing Pty Ltd*, in which the failure of a pipe in the defendant's waste treatment system caused the pollution of waters, contrary to section 120(2) of the PEOA 1997.[159] In addition to a $33,750 fine, the prosecutors requested that in light of the system failure, the defendant be required to submit to the authority a document specifying how the system was operated and how incidents were recorded. Legislation will also often empower a regulator to undertake restoration or prevention steps and recoup resulting costs from the offender.[160] However, as illustrated in Tables IX and X above, such orders are relatively rare in both NSW and Victoria.

[155] See press release available at: <http://www.environment-agency.gov.uk/news/99144.aspx> accessed 7 January 2009. Also see Producer Responsibility Obligations (Packaging Waste) Regulations 2007 (as amended) SI 2007/871.

[156] See eg EPA 1990 (Ontario) s 190 and Ontario Water Resources Act s 112, under which the court has very broad powers to issue a 'restoration order'. These orders require the offender, whether it be an individual or a corporation, to take action to 'prevent, eliminate or ameliorate damage that results from or is any way connected' to the offence. In England and Wales, where a defendant is convicted of committing an offence under the pollution prevention and control legislation, the courts have the power to award the equivalent of a 'clean-up order'. See Pollution Prevention and Control Regulations 2000, reg 35 . There would appear to be no equivalent provision in legislation relating to waste and water pollution.

[157] *EPA v Keogh* [1998] NSWLEC 225.

[158] S 245 PEOA 1997 (NSW).

[159] [2003] NSWLEC 82.

[160] See, eg, EPA 1970 (Victoria) s 66 and PEOA 1997 (NSW) ss 246 and 247.

iii 'Profit' and 'Forfeiture of Monetary Benefit' Orders

As we have seen, a fundamental principle in the sentencing of criminal defendants is that the offender should not profit from committing an offence and that the fine should aim at recovering any financial benefit accruing to the defendant. Under the deterrence framework which calculates penalties according to the utility/benefit accruing to the offender, removing any cost-savings or economic gain from non-compliance will not only act as a deterrence, but can also level the playing field for competing businesses that are complying with the law. This was recognised by Stuart CJ as early as 1980 in the Canadian case of *United Keno Hill Mines*, when he stated that:

> The assessment of a fine based on illegally obtained gains is essential to ensure that non-complying corporations do not acquire an economic advantage over complying competitors.[161]

In many jurisdictions sentencing considerations require the courts to ensure that profits accruing from the offence are reflected in the fine imposed. This is indeed the case in England and Wales, where the Sentencing Advisory Panel advice states that the courts must take into account 'any economic gain achieved by the offender by failure to take the appropriate precautions'.[162] But it would seem that mere guidance is not enough. Reviews, reports and consultations on environmental law enforcement in England and Wales have revealed a perception that when sanctioning environmental offenders, the English courts are not generally removing the financial gains made by the offender in committing the offence. This is, in effect 'giving businesses an incentive to continue to fail to comply in return for profit.'[163] The Macrory Review provides several examples of fines that fail to reflect the profit made by a firm in not complying with environmental regulation, the most revealing of which concerns a small waste disposal company that saved £250,000 by operating without a waste management licence. The company was fined just £25,000.[164] By virtue of this fact alone, the deterrent impact of the criminal justice system is extremely weak.

This problem may be overcome by giving the courts explicit power to award profits or monetary benefits penalty orders which, in most cases, are not subject to any maximum level.[165] For example, section 189 of the EPA 1990 (Ontario) states:

> The court that convicts a person of an offence under this Act, in addition to any other penalty imposed by the court, may increase a fine imposed upon the person by an amount equal to the amount of the monetary benefit acquired by or that accrued to the

[161] Above n 19, para 33.
[162] Above n 3, para 16.
[163] Above n 139 (Macrory Consultation) para 2.6.
[164] *Ibid.* See also DEFRA Review above n 139, 19–20 and DEFRA *Review of Enforcement: Review of Evidence*, 10–11, and P Hampton, *Reducing administrative burdens: Effective inspection and enforcement* (HM Treasury, March 2005) para 17.
[165] See, eg, PEOA 1997 (NSW) s 249(1). There is no equivalent power in Victoria.

person as a result of the commission of the offence, despite any maximum fine elsewhere provided.[166]

Of course, a profits order may not be appropriate in all circumstances. This is recognised by guidance in NSW which states that an order will be sought where the EPA can 'quantify the benefit obtained' and it is satisfied that 'the offender has sufficient funds to pay all or a significant proportion of the benefit obtained'.[167] There are some examples of such orders being used in Australian and Canadian environmental cases. Take the Canadian case of *R v Phoenix Powder Coating Inc*, in which the court, in addition to a $14,000 fine, imposed a penalty of $16,000 in order to deprive the defendant of the monetary benefit obtained when liquid waste was unlawfully disposed of in a ditch.[168]

Despite the theoretical arguments in favour of using the criminal justice system to recoup profits accrued in offending, prosecution reports in many states would suggest that very few courts exercise their powers in this regard. For example, as indicated in Table IX above, from 2003–07 the LEC in NSW did not serve any profit orders. Similarly, an examination of Ontario EPA litigation summary documents for 2003 and 2004, and a search of court reports for 2007–08 reveals that no monetary benefit orders were issued by the courts during this time. This is somewhat perplexing as on the face of it, profits orders appear to be extremely useful sentencing options. One possible explanation is that in the majority of cases, the profit accruing to the offender is not easily quantifiable.

As noted earlier, courts in England and Wales do not have the power to award a profits order. This is not to say that the removal of economic gain cannot be achieved through alternative means. Section B i above illustrated how a compensation order has been used to remove the savings made by a company in not paying licence fees, though examples are rare. A confiscation order can also be made under the Proceeds of Crime Act 2002. This power is, however, limited in its application[169] and has not been used successfully in environmental cases.[170] The failure of the courts to reflect economic gain has led to calls from various quarters for the introduction of profits orders. Macrory comes out in favour of profits orders where economic gain can be clearly calculated (for example, where licence fees have been avoided). Such a power would not, he argues, replace the need for courts to take into account any savings made when determining an appropriate fine.[171] A profit order would constitute 'a non-judgemental sanction [that would reflect] solely the profits made from non-compliance.'[172] This could lead to a large

[166] See also Environmental Protection and Enhancement Act (Alberta) s 230, Ontario Water Resources Act s 111 and CEPA s 290 ('additional fine').

[167] EPA (NSW), *Guidelines for seeking environmental court orders* available at <http://www.environment.nsw.gov.au/legislation/environ_courtorders.htm> accessed 5 January 2009.

[168] Unreported, Ontario Court (Provincial Division), 13 August 1991.

[169] Most importantly for our purposes, confiscation orders cannot be made by Magistrates' Courts, where the majority of environmental cases are heard (s 6).

[170] Above n 164 (Review of Evidence) para 3.24.

[171] Above n 139 (Macrory Report) paras 4.48–4.49.

[172] *Ibid* para 4.49.

profits order, where savings were high, but a small fine where, for example, the offence was committed carelessly and the environmental harm resulting from the offence was minimal. This, he concludes would:

> [S]trengthen enforcement and send a clearer signal to industry that it is not acceptable to make financial gain from non-compliance . . . provide a more level playing field . . . and provide a deterrent for non-compliant businesses.[173]

In short, if used appropriately, profit orders would undoubtedly strengthen the deterrent impact of the criminal law and should be seriously considered for introduction in England and Wales.

iv Payments for Environmental Purposes

By requiring an offender to inter alia contribute to a fund, scholarship or environmental programme, the sentencing options discussed in this section are clearly monetary sanctions that suffer from many of the limitations of criminal fines. However, by directing payment at environmental issues, the orders discussed below serve important retributive, reparative and rehabilitative functions.

In Canada, courts may have the power (either generally or explicitly) to require an individual or company to establish or contribute to a trust fund or scholarship that is used to support an environmental programme of some kind, or to make a payment for other general environmental purposes.[174] There are however very few examples of the provincial courts in Ontario using their powers for this purpose. This is not a pattern across all Canadian provinces with the courts in, for example, Alberta regularly demanding that environmental offenders make a financial contribution to a fund, scholarship or an environment-related project.[175] For example, in March 2007, ConocoPhillips Canada Resources Corp. was convicted of air pollution.[176] In addition to a $20,000 fine, the company was ordered to pay $200,000 in favour of the University of Saskatchewan for environmental research. Outside the pollution control area, at federal level there are some examples of such orders being awarded under the Fisheries Act. In *R v Corner Brook Pulp and Paper Ltd* the defendant was convicted of depositing or permitting the deposit of acutely lethal effluent into water contrary to section 36(3) of the Fisheries Act.[177] The company was fined $500,000, $200,000 of which was to be used to fund scholarships within educational institutions. $75,000 of this was allocated to the

[173] Above n 139 (Macrory Report) para 4.49.

[174] Explicit provision is made for such orders in CEPA (Canada) s 291(1)(p). General provision is made under EPA 1990 (Ontario) s 190(1)(c).

[175] Under the Environmental Protection and Enhancement Act (Alberta), the courts have a general power to require 'the offender to comply with any other conditions the court considers appropriate in the circumstances for securing the offender's good conduct and for preventing the offender from repeating the same offence or committing other offences.'

[176] See Alberta Environment, *Enforcement of the Environmental Protection and Enhancement Act* (1 January 2007–31 March 2007) 6.

[177] (1996) 22 CELR (NS) 199.

scholarship fund at West Viking College for students attending the Reserve Technology Program and the remaining $125,000 was awarded to a scholarship fund of the Sir Wilfred Grenfell College for students entering the BSc programme in Environmental Studies.[178]

In Australia, the Victorian courts have no such power, but a payment of money through an environmental service order (discussed in Section C below) arguably serves a very similar purpose. In NSW, section 250(1)(e) empowers the LEC to order the defendant to pay a sum of money to inter alia a specified organisation for general environmental purposes or for a specified project. The LEC has given a broad definition to the term 'environmental organisation' by including government departments. For example, in May 2006, Arenco Pty Ltd pleaded guilty to causing water pollution on two separate occasions.[179] The company was ordered to pay $26,000 to a government mining authority to be used for an erosion and sediment control project at an abandoned silver mine. The LEC has made it clear that orders relating to payments such as these should be made in conjunction with a publication order.[180]

Offenders may also be required to pay into legislatively established environmental funds.[181] In Canada, for example, Environment Canada administers an Environmental Damages Fund (EDF), money from which is allocated to projects which restore natural resources, usually in the same area where the pollution occurred. Although this is the funding priority, other projects such as research and development on environmental damage assessment and restoration and education on pollution prevention may be considered suitable for funding. Environment Canada allocates funding to eligible local organisations such as NGOs, universities, aboriginal groups and provincial, territorial and municipal governments. There are numerous examples of companies convicted of offences under CEPA or the Fisheries Act being ordered to pay into the EDF. For example, in April 2004, Metalex Products Ltd of British Columbia pleaded guilty to one count of unlawfully importing a hazardous waste, namely scrap lead shot, into Canada. The BC Provincial Court fined the company $500 and ordered it to pay $4,500 to the EDF. The money is to be used to assist in a study which will determine the cause of lead poisoning of trumpeter swans in the Fraser Valley.[182] Individuals can also be required to pay into the EDF. For example, in August 2001, a settlement was released in the prosecution of Carl Hunt. As part of the settlement, Mr Hunt was required to pay $4,000 into the EDF. An EDF award can also be made where a company is convicted of an offence under provincial as opposed to federal legislation. This is illustrated by a case brought by Alberta Environment

[178] A further $50,000 was allocated to the Corner Brook Development Corporation dedicated to projects relating to the environmental improvement of the Corner Brook stream area.

[179] *Environment Protection Authority v Arenco Pty Ltd* [2006] NSWLEC 244.

[180] *Environmental Protection Authority v Centennial Newstan Pty Ltd* [2006] NSWLEC 732, [141], cited in Preston above n 18, 163.

[181] See also, eg, the Environment Trust (NSW) established under the Environmental Trust Act 1998 (s 250(1)(e) PEOA 1997).

[182] Environment Canada News Release.

in 2003. In July that year Canadian 88 Energy Corporation pleaded guilty to unlawfully releasing or permitting the release of a substance into the environment which causes or may cause a significant adverse effect. The company was fined $78,000, $67,500 of which was to be paid into the EDF for the purpose of carrying out a creative sentencing order.[183]

C Publicity Sanctions

i Introduction

The extent to which adverse publicity will have a deterrent effect on corporate offending in particular has been a subject of debate for many years with scholars investigating, both theoretically and empirically, the stigmatising effect of negative publicity on individuals and corporations. Although publicity sanctions have become more prevalent in recent years, there are examples of statutory provisions providing for the publication of convictions dating back to the 19th century.[184] Many relate to food, drink and drugs law. For example in the UK, the Bread Act of 1822 provided that:

> It shall be lawful for the Magistrate or Magistrates, Justice or Justices, before whom any such Offender or Offenders shall be convicted, to cause the Offender's Name, Place of Abode and Offence, to be published in some Newspaper which shall be printed or published in or near the City of London or the Liberty of Westminster, and to defray the Expence of publishing the same out of the Money to be forfeited—in case any shall be so forfeited as last mentioned, paid or recovered.[185]

Similar examples can be found in Australian law where, in addition to publishing in the written word, there is one example of a statute which allowed for publication by radio broadcast.[186]

In this section, I consider formal as opposed to informal publicity. Informal publicity, by which I mean the reporting of charges, hearings, convictions and sentences by the mass media at their own discretion has been utilised globally for some time, in many different situations, and will undoubtedly remain an important informal enforcement tool for decades to come. Its impact should also potentially be taken into account in determining the quantum of sanction imposed by a court or administrator. But formal publicity, that is publicity 'which is activated for the purpose of imposing a sanction either by the court or administrative body itself, or by some other official agency'[187] and its impact on corporate crime is the focus of this section. The idea has been embraced in both Australia and Canada and it is likely that in the foreseeable future, the English courts will be vested with

[183] Alberta Environment, *Enforcement Actions 1 July–30 September 2003*, 10.

[184] For a detailed overview, see B Fisse, 'The Use of Publicity as a Criminal Sanction Against Business Corporations' (1971) 8 *Melbourne University Law Review* 107, 110–17.

[185] See 6 & 7 Will IV c 37, ss 6 and 14.

[186] Black Marketing Act 1942 s 13(1)(a), (b).

[187] Above n 184, 109.

the power to publicise a polluting offence. Before examining the legislative provisions in detail, it is interesting to consider, more generally, the deterrent impact of formal publicity and its use as a judicial sanction. This section begins by critically analysing the strengths and weaknesses of publicity sanctions, and considering important factors to consider in designing adverse publicity provisions. It concludes by examining adverse publicity provisions available to courts in Australia, Canada, and England and Wales when sanctioning polluters.

ii Strengths

The precise nature and form of a formal publicity sanction is largely determined by what is to be achieved. The main claim of those who advocate the use of adverse publicity is that it has important effects for deterrence. Fisse notes three primary deterrent functions of publicity: the infliction of monetary loss, the loss of prestige and respect and its ability to induce government intervention.[188] All three can strengthen the deterrent value of the criminal sanction (D) in the deterrence framework. That publicity sanctions will inflict financial loss on the corporate offender is not doubted. Placing an advertisement about an offence in a newspaper may influence consumers and lead to a drop in sales, reduced profitability or a reduction in share value. It is estimated that a company's image accounts for up to four per cent of its stock price.[189] However, if the deterrent value of the sanction is solely financial, then it is extremely weak and the same monetary loss could be achieved by increasing fines.

What about the impact of publicity on corporate prestige and respect: is it true to say that 'even the wealthy can wilt from social disapproval'?[190] Fisse and Braithwaite, in their leading study of this area, found that the companies who participated in their investigation deemed adverse publicity to be of concern, 'not so much by reason of its financial impact but because of a variety of non-financial effects, the most important of which is corporate prestige.'[191] The companies were sensitive about prestige as an interest, over and above profits. Similar sentiments are echoed by Hawkins in Environment and Enforcement:

> Few officers suggest that there is a stigma with economic implications reflected in damage to sales and profitability which attaches to a manufacturer found guilty of polluting a watercourse. Instead, it is assumed that a company will seek to protect its reputation as a good-in-itself.[192]

[188] *Ibid* 117–126.

[189] B Fisse and J Braithwaite, *The Impact of Publicity on Corporate Offenders* (Albany, State University of New York Press, 1983) 248.

[190] Above n 184, 118.

[191] Above n 189, 247. The book uses, as case studies, 17 large transnational companies which suffered adverse publicity crises in the 1960s. Although they do not deal with formal publicity as defined above, they make some very interesting observations about publicity and corporate deviance.

[192] K Hawkins, *Environment and Enforcement: Regulation and the Social Definition of Pollution* (Oxford, Clarendon Press, 1984) 116.

In summarising the findings from the case-studies, Fisse and Braithwaite conclude that the 'financial impacts of any significance occurred in only a small minority of case studies'.[193] On the other hand, non financial impacts such as 'loss of corporate and individual prestige, decline in morale, distraction from getting on with the job, and humiliation in the witness box' were acutely felt.[194] This, coupled with the fact that a publicity sanction is not limited by the amount of money which a company can afford to pay, would suggest that it may have a greater deterrent effect that traditional monetary fines. This is arguably of particular importance in an area of crime dominated by white-collar offenders.

A third value inherent in publicity is the fact that it may induce government intervention. This is not to suggest that the criminal courts should force action by government bodies and quasi-regulatory authorities. Rather, that publicity would bring to their attention the fact that a company had been convicted of an offence. For example, it may intensify investigative action, trigger the introduction of new legislation or even instigate a formal inquiry. As regulatory bodies may be alerted to the offending company, this potentially increases compliance with other regulatory regimes.

Publicity sanctions can also serve several other important functions. Fisse and Braithwaite identify the fact that 'in every case, there was some worthwhile reform.'[195] Many pollution offences are the result of inadequate working practices and ineffective management systems and so if publicity sanctions can prompt procedural reform then this would surely be desirable. Of course, such reform is not certain.[196] But despite the fact that procedural/policy reform is not guaranteed, the mere possibility of reform provides, in my view, a strong argument in favour of adverse publicity. The NSW Law Reform Commission has also recognised the fact that publicity sanctions reflect the expressive/retributive dimension of the criminal law, in that social censure is an important element of criminal punishment.[197] This can be contrasted with a monetary fine which, it can be argued, conveys the message that: (a) corporate crime is less serious than other crimes; and (b) companies can buy their way out of trouble. Fisse also notes three supplementary advantages of a publicity sanction: its general deterrent impact (in the sense that publicity will increase the chances of other companies operating in the same regulatory environment being alerted to a successful criminal convention), its ability to inform the public about the operation of relevant legislation and its ability to warn buyers of defective products, although the latter is arguably less relevant in the context of pollution control regulation.

[193] Above n 189, 243.

[194] *Ibid*

[195] *Ibid.* These are just some of the conclusions reached by Fisse and Braithwaite.

[196] As noted by the Australian Law Reform Commission: '[P]ublicity orders cannot guarantee reform of the procedures or policies that contributed to or caused the commission of the offence, [but] they could be used to put public pressure on the corporation to carry out such reforms.' ALRC Issues Paper 20, 2001, *Sentencing: Corporate Offenders*, ch 3 para 3.21.

[197] Above n 76, ch 11, para 11.2.

iii Weaknesses

But publicity sanctions have not escaped extensive criticism with Coffee and others questioning whether adverse publicity is actually effective as a corporate sanction.[198] One disadvantage of publicity as a criminal sanction is its uncertainty. Whilst variations in the wealth of the offender can be relatively easy to take into account when determining an appropriate fine by, for example, setting fines as a percentage of corporate turnover, it is less easy to determine the financial (and non-financial) impact of a publicity sanction. Another complicating factor is that a company which values its corporate reputation will be more affected by publicity than one which sees all publicity as good publicity. Moreover, the precise impact of publicity (and therefore the quantum of the sanction) depends not on the regulator or the court but on the 'capricious jury of public or governmental opinion.'[199] Coffee, in discussing financial impact (in other words the negative effect of publicity on the profitability of the company) argues that, like monetary fines, this may create spillover effects which are felt by employees, shareholders and consumers rather than key decision-makers within the corporation.[200] The impact of the sanction may therefore be felt disproportionately by innocent workers, suppliers and distributors. However, if one is to accept that the aim of publicity is to impose a non-financial as opposed to a financial loss, these particular criticisms hold less sway for spillover effects will be minimal and the uncertainty of the sanction is less problematic. Fisse and Braithwaite's study revealed that serious financial spillovers are likely to be the exception rather than the rule and where they were likely to occur, 'there was no evidence that the financial impact would fall disproportionately on workers at the bottom of the hierarchy.'[201]

A second disadvantage is that a company can avoid the impact of a publicity sanction by, for example, dissolving, re-branding its products or even changing its name. But one of the most probable and least expensive evasion tactics is the use of counter-publicity. This is likely to dilute the effectiveness of any publicity sanction, particularly if one accepts the proposition that the government is a relatively poor propagandist. Unlike a court or regulatory agency, companies have at their disposal a range of advertising techniques such as catchy slogans or phrases which, according to Fisse, would be rejected by official agencies on the grounds of distortion or vulgarity.[202] In other words, the public are likely to be more affected by such tactics than an official statement of the type normally required by a court. Furthermore, the company is in a position to launch a counter-publicity campaign in advance of any formal publicity sanction.

However, even if a firm does launch a counter-publicity campaign, publicity may remain a desirable sentencing tool. Evidence would seem to suggest that

[198] Above n 35.
[199] Above n 184, 140.
[200] 'Spillover' effects were discussed in s III C.
[201] Above n 76, para 11.20.
[202] Above n 184, 136.

counter-publicity is only used where the publicity sanction is severe. And it is worth bearing in mind that if one supports the proposition that the primary rationale behind publicity is to inflict a monetary loss on the offender, the cost of counter-publicity could, in certain circumstances, be greater than a monetary fine. The concerns over counter-publicity could be tackled by using the sanction itself to prohibit or limit the use of counter-publicity or even advertising, which, like counter-publicity can be used by a firm to regain public confidence in its products or services. Although this has been mooted in academic circles, there would appear to be serious constitutional and rights-based objections.[203]

If the purpose of formal publicity is either to lower corporate prestige in the eyes of the general public or to inflict a large financial loss on a company by persuading consumers and others not to buy its products, a third main disadvantage of publicity could relate to problems of persuasion. In the case of publicity associated with pollution, adverse publicity must compete with all the other factors which contribute to a company's reputation including turnover, price and quality of goods or services. And finding an acceptable way of lowering reputation is difficult. Formal publicity which notifies the public that a company has been convicted of an offence will only be effective if the public understands what they should do with the information. According to Fisse, this is the main obstacle in finding an effective method of persuasion.[204] There are also problems inherent in the nature of corporate criminal responsibility and regulatory offences. As we have seen, corporations are normally convicted of strict liability crimes.[205] The fact that a company can be found guilty of an offence even if they have exercised due care and attention could result in public sympathy rather than antipathy. And many regulatory, as opposed to 'traditional' crimes are not known or understood by the public. Of course, discretion in prosecutorial decision-making may well limit prosecutions to those cases where failure to exercise reasonable care (or some form of mens rea) can be proven. Nevertheless, it may be necessary to appeal to the self-interest of the public by appealing to the reason of the people. Bentham uses the example of smuggling to illustrate how publicity could be used effectively:

> Whosoever deals with smugglers let him be infamous. He who buys uncustomed goods, defrauds the public of the value of the duty. By him the public purse suffers as much as if he had stolen the same sum out of the public treasury. He who defrauds the public purse, defrauds every member of the community.[206]

In summary, if formal publicity is to have the desired effect, it is important that the public can relate to the offence committed.

So, overall, does formal publicity perform an important deterrent function? In my view, there is no doubt that it inflicts an important financial loss and has some distinct advantages over monetary fines. But although this monetary loss is inevitable, I would argue that the real deterrent value of the sanction is not finan-

203 For a summary of the objections see Fisse *Ibid* 138.
204 For details, see Fisse *Ibid* 128–30.
205 See further ch 6 s IV.
206 Bowring (ed), *The Works of Jeremy Bentham*, 465.

cial but is loss of corporate reputation or prestige. If one accepts the proposition that adverse publicity stigmatises corporations and individuals (and imposes a financial loss), the issuing of a formal publicity order will raise the 'costs' to the offender (the value of D in the deterrence framework) and will therefore increase the deterrence value of prosecution. This is not to say that formal publicity is appropriate as a sanction in all prosecutions. The effects of adverse publicity are potentially extremely large, and in any event, routine publication of convictions may have a desensitising effect on the public and orders may subsequently lose some of their effectiveness. The use of formal publicity should therefore be reserved for the more serious cases where for example the offence involves a serious pollution incident, where the company has a poor record of compliance or where a monetary penalty, which has been reduced by the court because of the company's financial situation, could be seen as trivialising the offence.

iv Designing a Sanction

If lowering prestige is indeed the primary purpose of formal publicity, then how might a publicity sanction be designed? Should the sanction, for example, target individual directors or employees as well as the company itself? Although a sanction directed at the company will have some effect on the personal reputation of directors and superior officers, its effect will be much less than a sanction which identifies individuals. The central tenet of Coffee's paper (that prosecution should be targeted at both the corporate offender and the individual company executives) leads him to support the use of publicity sanctions which identify the culpable individuals after the corporation has been convicted. As identifying and convicting individuals within a firm can be difficult, a publicity sanction targeted at the individual 'may in many instances be the only available way to censure the culpable manager.'[207] This may be acceptable if the relevant individuals have been located and convicted of the offence. But where those persons have not been identified and convicted, then there are serious objections to such an approach. This leads Fisse to conclude that in such circumstances, a fine may be a more appropriate sanction.

Moreover, at whom should the publicity be directed at? Who is intended to be influenced by the sanction? Bearing in mind our discussion on 'problems of persuasion' a case can be made for targeting high level business executives and opinion leaders. Such individuals would be more aware of the nature of corporate crime and regulatory offences, and would be less likely than the public generally to be swayed by emotive slogans and counter-publicity. The intended audience is obviously determined by identifying what it is that the publicity sanction is set to achieve. Publicising the offence more widely in, for example, the national or local press, may not necessarily be desirable but where the sanction is aimed at

[207] Above n 35 (Coffee), 433.

inflicting loss of reputation amongst consumers and the public, then wide-spread publication may be appropriate. None of the suggestions above detracts from the fact that the main deficiency of publicity sanctions is uncertainty. However, if the sanction is targeted at a specific audience, and is used in conjunction with a monetary fine to inflict a definite monetary loss, the difficulty of uncertainty will be at least minimised.

v Publicity Sanctions in Australia

Many Australian states utilise formal publicity to deter environmental crime and in both Victoria and NSW, the courts have the power to make a 'publicity order.'[208] Whilst parliamentary support for the introduction of such orders was strong, it was not overwhelmingly so. For example, during the second reading of the Environment Protection (Enforcement and Penalties) Bill in the Victorian Parliament, the Shadow Minister for the Environment, in response to the proposal to introduce publicity requirements, went so far as to say:

> [But the other provisions] enforce publicity and self-mortification—almost a modern version of the stocks, with the medieval idea that a convicted person, a wrongdoer, ought to be laid out in stocks in a public place and held up to ridicule; that is an ancient notion.[209]

Nevertheless, since their introduction both the courts and regulators appear to support the view that publicity sanctions increase awareness, improve deterrence and hold offenders to account. For example, according to Justice Preston sitting in the LEC:

> Publicising sentences for environmental crime improves the effectiveness of sentences as a deterrent. This is particularly applicable to corporate offenders, who are susceptible to criminal stigma.[210]

In addition, correspondence with regulatory and legal officers in both Victoria and NSW expressed their support for publicity orders.[211]

When a publicity order is awarded, the offending individual or company must publicise the offence, its environmental or other consequences and the penalties and other orders imposed as a result of the commission of the offence.[212] With a view to targeting the adverse publicity at a particular audience, the legislation also

[208] In Victoria, see EPA 1970 s 67AC(2)(a), (b), as inserted by Environment Protection (Enforcement and Penalties) Act 2000 s 5. In NSW see PEOA 1997 s 250(a), (b). In NSW, the power to make a publicity order can only be exercised by the LEC.

[209] Per Mr Perton (Doncaster) Second Reading in Victorian Assembly 31 May 2000, 2051.

[210] *Environment Protection Authority v Waste Recycling and Processing Corporation* [2006] NSWLEC 416, [242].

[211] For example during an interview with the Director of Legal Services at the EPA NSW in May 2004, it was stated that publicity orders 'hurt industry more than fines' and can be 'a very successful deterrence.' Consequently, 'industry is very resistant to the use of them.'

[212] See EPA 1970 (Victoria) s 67AC(2)(a), (b), and PEOA 1997 (NSW) s 250(1)(a), (b).

gives the courts discretion in determining when and where the publicity notice must be placed. Where the publicity is to be targeted at the public generally, a notice may, for example, be placed in state or national newspapers.[213] Alternatively, the notice could be directed at a specified class of persons (such as shareholders through company annual reports) or aggrieved individuals through notices to neighbours and local communities. The court can even go so far as to specify the exact wording to be used, the layout of the piece and the size of font.[214] The prosecutor will normally submit the precise wording of the order for consideration by the court.

(a) be in the wording specified [below] and must not be supplemented by additional text;
(b) be published within 30 days of the date of this Court Order;
(c) be publicised in the Early General News section of the following papers;:

    i. Whitehorse Leader
    ii. Progress Leader
    iii. Financial Review
    iv. Herald Sun
    v. The Age

(d) be of a minimum size of 12cm by 3 columns
(e) contain the Defendants co rporate logo

Figure XII: Example of Publicity Order, Victoria, Australia[215]

Publicity orders are however only appropriate in certain circumstances. Guidance produced by the EPA NSW on creating sentencing options states that a publication order is normally sought against corporate offenders, where the deterrent effect of fines is limited.[216] In deciding whether or not to seek a publication order, the EPA will take into account a number of factors including the defendant's culpability and prior record and the environmental harm caused by the offence. Irrelevant factors include co-operation during the investigation and the entering of a guilty plea.

As indicated by Tables IX and X above, judges in both Victoria and NSW have been receptive to their powers. Examples provided in EPA (Victoria) Annual Reports indicate the extent to which the courts prescribe the nature and form that

[213] See EPA 1970 (Victoria) s 67A(2)(a) and PEOA 1997 (NSW) s 250(1)(b).

[214] See Figure XII below.

[215] This is an example of a publicity notice imposed on Amcor Packaging (Australia) Ptd Ltd. The company was convicted of polluting waters contrary to s 39(1)(c) EPA 1970 (Victoria). EPA Victoria, *Annual Report 2005–06* App.

[216] Above n 167, 6.

the publicity order must take, with many cases requiring publication in local, state and national newspapers including the Herald Sun, The Age and The Australian. The NSW LEC has also imposed publicity orders in several recent cases. In 2005–06 for example, Arenco Ptd Ltd was successfully prosecuted by the EPA for polluting waters contrary to section 120(1) of the POEO Act 1997.[217] In addition to funding an erosion and sediment control project, the company was required to publish details of the offence in The Sydney Morning Herald and The Western Advocate.

vi Publicity Sanctions in Canada

Publication or adverse publicity orders are also commonly provided for in federal and provincial Canadian environmental statutes. For example, Alberta's Environmental Protection and Enhancement Act permits the courts to make an order 'directing the offender to publish, in the prescribed manner and at the offender's cost, the facts relating to the conviction.'[218] Where an offender fails to comply with a publication order, the minister is commonly entitled to publish the facts in compliance with the order and recover the costs from the offender.[219] Similarly, section 291(1)(g) of the CEPA allows the courts to direct an offender to publish, in the manner directed by the court, the facts relating to the conviction. Interestingly, Ontario's environmental protection is notable for not empowering the courts with such a power, although similar powers exist under the (general) Provincial Offences Act. Under section 72(1)(b) of that Act, a defendant can be required to comply with the conditions of a probation order, which can include an adverse publicity requirement. This provision was used when Bata Industries, convicted of permitting the discharge of liquid industrial waste from its shoe manufacturing process onto the ground, was required to: (a) publicise the facts and circumstances leading to the commission of the offence in a newsletter; and (b) publicise the proper standards of toxic storage in a technical circular.[220]

The wording of these orders is at the discretion of the courts, who are often guided by prosecutors. The following publication order illustrates that the courts can be quite creative in exercising their discretion. Harjinder Johal and RJS Investments inc were convicted of accepting empty beverage containers from outside Alberta, contrary to Alberta recycling regulations. The court ordered the company to place an advertisement in the Edmonton Journal. The advert began 'I GOT CAUGHT. I HAVE BEEN CHARGED AND CONVICTED OF ACCEPTING EMPTY BEVERAGE CONTAINERS FROM OUTSIDE ALBERTA'. To put icing on the cake, the defendants were ordered to post a sign by the cash register stating that 'it is illegal to accept containers from outside Alberta'!

[217] *EPA v Arenco Pty Ltd* [2006] NSWLEC 244.

[218] S 234(1)(c). See also Environment Act (Nova Scotia) s 166(1)(c).

[219] See, eg, CEPA s 291(2) and Environmental Management Act (British Columbia) s 127(2).

[220] See *R v Bata Industries Ltd* above n 115. The decision of the trial judge to require publication worldwide was quashed on appeal, and the publication requirements were limited to Canada.

It would seem that, compared with their counterparts in Australia, the Canadian regulators are less likely to request a publication order and the Canadian courts are much less likely to grant one. A search of the Ministry of Environment (Ontario) court bulletins reveals that no publication orders have been issued in recent years. It would seem that these orders are used most frequently in the province of Alberta.

vii Publicity Sanctions in England and Wales

In England and Wales, the Environment Agency clearly recognises the deterrent impact of adverse publicity as press releases of significant prosecutions are sent to all relevant media including local, regional and national press, regional TV and radio and environmental publication.[221] But at present, the English criminal courts cannot require an offender, as part of the sentencing process, to publicise an offence. If introduced, the offender (as opposed to the regulator) would bear the costs of promoting the offence, and, in contrast with existing practice, an independent third party (in this case, the courts) would ensure that the offender has received a fair and objective assessment of the offence. The Environment Agency itself has recently advocated the introduction of formal adverse publicity orders whereby 'a court could make an order that a notice, (with wording agreed by the Agency as to the offence and its circumstances), be placed in the local or even national media within a specific time.'[222] The DEFRA and Macrory Reviews have also recognised the financial and non-financial consequences of reputational sanctions and have both come out in favour of giving the criminal courts the power to impose a publicity order, in addition to or instead of any other sentence.[223] All in all, there is clear impetus for reforming the existing system of Environment Agency-led adverse publicity. Of course, guidance should be drafted to reflect the circumstances in which publicity orders should be requested and awarded. In this vein, DEFRA's concluding report suggests that adverse publicity orders should possibly be limited to cases involving serious culpability, without which they may seem 'disproportionate and oppressive.'[224]

[221] And in a recent study, Environment Agency officers stated that adverse publicity was the most important consequent of prosecution, particularly when applied to large-scale industry which has a reputation to damage. See P de Prez, 'Beyond Judicial Sanctions: The Negative Impact of Conviction for Environmental Offences' (2000) 2 *Environmental Law Review* 11.

[222] Supplementary memorandum from the Environment Agency presented to the Environmental Audit Committee, at 7(c). Available at <http://www.publications.parliament.uk/pa/cm200304/cmselect/cmenvaud/126/4012208.htm> accessed 5 January 2009.

[223] Macrory Report above n 139, para 4.69. DEFRA Review above 139, para 6.3.51.

[224] *Ibid* (DEFRA Review).

D Corporate Community Service

i Introduction

In 1983, a group of six bakeries in the United States were successfully prosecuted and fined for price fixing. The court, however, excused the bakeries from paying a substantial proportion of the fine on the condition that they provide, for a 12-month period, a free supply of freshly baked products to a number of needy organisations in the New York City area.[225] In doing so, there were two key factors which influenced the court. First, a fine which reflected the seriousness of the offence would have led to bankruptcy in all six cases, thereby putting the bakeries out of business. And second, by substituting the payment of the fine with a community project such as this, the court was requiring the defendant companies to make "symbolic restitution' for their offences by doing something more exacting and thought provoking than merely paying the fines.'[226] By using community service as a corporate criminal sanction, the court had found an innovative and effective means of sentencing corporate offenders.

For some time now, the criminal courts, in many jurisdictions, have, as we have seen, been vested with the power to sentence a convicted individual to perform community service by undertaking socially useful work.[227] With its emphasis on reparation, rehabilitation and retribution, it confers benefits on both the offender and society and is often regarded as a more effective sanction than imprisonment. Coupled with the associated reduction in prison costs, it is now seen as one of the major sentencing tools of the criminal courts. Corporate community service is, in contrast, a relatively novel means of tackling corporate crime.[228] However, it is gaining in popularity. In the environmental context, some state and provincial legislatures in Australia and Canada have recently introduced 'environmental project orders' which, generally speaking, require the offender to carry out a specified project for the restoration or enhancement of the environment in a public place or for the public benefit. Unlike community penalty provisions currently available in England and Wales, such orders can be served on both individual and corporate offenders. Furthermore, they ensure that any community service is focused on the environment. Before examining these environmental project orders in more detail, we will begin by considering, more generally, the value of community service, particularly when used against corporate entities.

[225] *United States v Danilow Pastry Co* 563 F Supp 1159 (SDNY 1983).
[226] Fisse above n 35, 245.
[227] Refer to s IV B for more detailed discussion.
[228] However, in the United States community service has been used in some cases as a condition of corporate probation, mitigation of sentence or non-prosecution since the 1970s. For details see B Fisse, 'Community Service as a Sanction Against Corporations' (1981) *Wisconsin Law Review* 970, 971–8.

ii Strengths

Increased deterrence (both general and individual) is just one of the advantages community service has over monetary fines. As already noted, fines are limited by what is known as the 'deterrence trap' whereby the maximum fine which can realistically be imposed is limited by the liquidity of the offending company: a court will not normally set a fine so high as to bankrupt the company. By ordering community service, the company may have to render a service or use non-cash resources as opposed to spending liquid assets. This allows the court to set a sentence which is commensurate to the seriousness of the offence. Of course, in many cases, community service will require some form of direct monetary contribution and if, for example, the order requires the company to make a financial contribution to a charity or other project, then the deterrence trap may not be avoided. The point is that community service offers flexibility, whereas fines do not. In addition, community service may, depending on its precise terms, require the company to allocate personnel to the running of the project. By doing so, it acts as a strong catalyst for internal discipline and highlights the seriousness of the crime to those officers and employees who were involved in the commission of the offence. As stated by Fisse:

> In being forced to allocate personnel to a required project or community service, a corporate offender would be encouraged to ask those responsible for subjecting it to that requirement to extricate the organization from the problem they created.[229]

Furthermore, if a company is sentenced to community service, the offence is more likely to be brought to the attention of the public than if prosecution results solely in a monetary fine. This, in itself, may be of some deterrence value. So, according to the deterrence framework, if D is corporate community service instead of (or as well as) a monetary fine, the deterrence value of proceeding to the prosecution stage is increased, the costs of non-compliance are more likely to outweigh the benefits and therefore the offending company is more likely to comply with it's statutory obligations.

It is also important to recognise that community service reflects a number of other penological principles, thus giving it several advantages over monetary fines. It can be seen as a positive instrument that provides a means of restoration (or compensation): it can lead to reparation of the harm caused by the offence and could benefit the community more generally. Such sentiments were echoed by an Australian High Court judge who, in requiring a corporate offender to undertake an environmental project, commented that:

> An important factor in prosecutions of this kind is not so much a monetary penalty but the possibility of making orders that have the effect of enhancing the environment and encouraging its protection, and the avoidance of its degradation.[230]

229 *Ibid* 1004.
230 *EPA v Simplot Australia Pty Ltd* [2001] NSWLEC 264 [21] (Justice Pearlman).

Presuming that a community service order is appropriately designed, it is also compatible with any theory of retributive justice because it satisfies 'by requital or redemption, the common retributionist requirement that crime by annulled.'[231] It is also arguable that such an order has rehabilitative effects. By pressing home to appropriate personnel the unwantedness of corporate crime and the damage it may cause, community service may influence the attitudes and opinions of those responsible for designing and managing internal operating procedures. In such a way, community service may lead indirectly to a revision in company systems in order to avoid recurrence of the offence.

iii Weaknesses

So, all in all, corporate community service is superior to monetary fines in many ways. It targets not only the company and affected members of the public, but also, if appropriate personnel are nominated to complete the project, targets the blameworthy actors. Monetary fines, however formulated, are unable to target this final category. This is not to say that corporate community service is unproblematic for there are a number of difficulties associated with its application in practice, some of which can be ameliorated if the legislative provisions providing corporate community service are appropriately drafted, the courts are careful in utilising their power and community service is ordered in conjunction with other sentencing tools. One of the serious drawbacks for our purposes is the fact that the costs of administering corporate community service could be high, particularly if the company is an 'unwilling' participant in the designated project. Supervision and verification of compliance with the order, particularly if the task is entrusted to an external and independent body, may be costly but would counter the risk that companies may falsify compliance reports. This is not, however, an unsolvable problem. First, it is likely that in most cases, the company itself will be responsible for providing ongoing compliance reports and final verification of completion (to the satisfaction of the court or other regulatory body). And second, if an external body is required, any administration costs can be passed on to the offending company. So, in achieving deterrence, there is no reason why community service cannot be as costless as monetary fines in terms of administration.

It is also of some concern that a company would allocate innocent employees the task of complying with the community service order. If this were the case, such individuals could be victimised and any monetary loss would be passed on to innocent parties. However, an appropriately worded order could require the company to identify senior managerial and other staff that were 'responsible' for the offence. A further serious criticism of community service is that it is too lenient a sanction to be effective and it is not as serious a sentence as imprisonment. This argument can be countered in two ways. Imprisonment is never going to be an option for companies and, as we have already seen, it is extremely difficult to

[231] Above n 228, 1005

convict individuals within a corporation.[232] In any event, other analogues of imprisonment such as temporary (or permanent) closure of business or even advertising bans would provide weaker sanctions than community service because there are too many negative side-effects. There is also a danger that an offending company may use the community 'project' to boost its corporate reputation and image. However, a court could minimise such a risk by, for example, imposing community service in conjunction with formal publicity, or by ordering the company to refrain from referring to the project without reference to the criminal proceedings that led to the sanction. Another potential drawback of community service is that, by spending money on a socially beneficial project, funds and assets may be diverted away from fixing the problem that led to the original offence. But again, by combining community service with probationary, preventative or audit orders, this problem can be addressed by the courts from the outset. One final concern relates to the risk that the judiciary could be seen to use their sentencing powers indiscriminately by supporting pet projects or charities. However, by requiring that the project must be reasonably related to the offence, or even connected to a programme of action that has been authorised by regulation as suitable for community service, the risk could be avoided. Alternatively, the company could bear the expense of designing an appropriate project which would then be approved by the regulator and/or the court.

iv Designing the Sanction

Bearing in mind the benefits and drawbacks discussed above, it is useful to summarise how community service as a corporate sanction should be designed, both legislatively and in practice. Although the courts (and regulators) should retain considerable discretion, it is obviously important to ensure that the sanction is used in such a way so as to give effect to the aims of corporate criminal law. The first key point to make in this respect is that community service should be viewed as a punitive order (although of course it could also be considered as remedial depending on its purposes and conditions). This would ensure that compliance costs are not tax deductible and would also stress the punishment dimension of the sanction. It has been suggested that the consent of the corporate offender should be required. Although this may minimise administrative costs, community service should not be seen as any different from other sanctions such as remediation and revocation orders. If consent was required, companies could use this as an excuse for merely paying a fine. In many ways, this defeats one of the key advantages of community service over fines—its ability to avoid the deterrence trap. One point not considered so far is how much should a company be required to do under a community service order? In terms of legal certainty, it is obviously important that the 'amount' of community service is quantified by the court,

[232] See ch 6 s IV.

usually in monetary terms. The fine maxima could apply to the maximum for community service, although I would argue that the maximum should be higher. Undoubtedly, the court should set a time limit for completion of the project, otherwise a company could delay completion indefinitely. As we have seen, there should be some limits placed on the type of projects suitable for corporate community service, and the court (or regulator) should specify which corporate personnel should be involved in the project. Finally, administrative costs in terms of monitoring and final verification should be borne by the offending company but this would be the exception rather than the rule.

In conclusion, it can be strongly argued that in deterrence terms alone, community service is superior to a monetary fine. But the strength of the sanction is also in its flexibility and its ability to achieve many of the other aims of corporate criminal law. Although there are undoubtedly a number of drawbacks, many can be dealt with by careful design and implementation and they do not, in my opinion, detract from the strength of community service as a corporate sanction. As mentioned previously, some courts in Australia in Canada can not only sentence corporate offenders to community service, but can also target that community service to the environment. These environmental project orders are proving quite popular in Australia. At Canadian federal level, and in the province of Ontario there is, however, no evidence that community service orders have been used. I therefore focus in the following section, on the and availability and use of these orders in Australia.

v Environmental Service Orders in Australia

Under section 67AC(2)(c) of the EPA 1970 (Victoria) the courts can require offenders:

> [T]o carry out a specified project for the restoration or enhancement of the environment in a public place or for the public benefit (even if the project is unrelated to the offence).

The PEOA 1997 (NSW) contains a similar provision.[233] By imposing costs on offenders, these service or project orders clearly serve a deterrence function. They can also be seen as a positive instrument as, if used appropriately, they can generate environmental benefits to the community affected by the offence. They therefore reflect restorative and reparative values.

As noted previously, environmental service orders (ESOs) will not necessarily be appropriate in all prosecutions and there are clearly important factors to take into account in determining the suitability of this penalty in any particular case. Such factors should be borne in mind by both regulators (when requesting such orders) and the courts (in exercising their discretion in favour of an ESO award). Many of these considerations are reflected in the NSW EPA's guidelines on seek-

[233] S 250(1)(c).

ing ESOs.[234] First, an ESO may be suitable in lieu of a fine where, for example, due to the defendant's limited financial means, a fine would not reflect the seriousness of the offence. Second, an ESO will only be appropriate either in addition to or instead of a fine where the regulator can identify suitable projects in the locality in which the offence was committed; the project, of course, should not relate to work which the defendant would be expected to do in the absence of a criminal sanction to that effect. And finally, in order to minimise administrative costs associated with monitoring the progress of the project work, it is vital that the offender has the means, capability and willingness to undertake the work. As noted by Justice Pearlman:

> I think it is appropriate to make such orders in a case such as this where the defendant has signalled its willingness to be bound to orders of the nature of those specified in section 250(1)(c).[235]

An ESO will invariably be used with a publicity order because it is obviously important for the community to know that the offender is carrying out the work, not just as a gesture of good will but as a result of committing an environmental offence. Moreover, the courts often stipulate that the offender is prohibited from referring to the project without reference to the criminal proceedings.

This sentencing power is used most commonly by the Victorian courts, where nine out of the 13 major prosecutions brought by the EPA Victoria in 2006–07 led to a project order.[236] Interestingly, all of the ESOs issued in 2006–07 required the offender to pay a sum of money to a third party, who then undertook the work on their behalf. For example, in 2006, Waste Management Pacific Pty Ltd was required to pay $30,000 to Merri Creek Management Committee Inc, the money to be used to undertake environment restoration work and stormwater education in the local area.[237] Projects involving third parties also commonly require the money to be used, inter alia, for informational and educational purposes. For instance, in 2005, Ringwood Magistrates' Court in Victoria required Amcor Packaging (Australia) Pty Ltd to pay the sum of $60,000 to Help for Wildlife Inc, the money to be used to build 'a single all-purpose building which will perform the function of a wildlife education, rehabilitation and resource information centre.'[238] If the courts use their power to require the payment of a sum of money to a specified organisation (as they do in Victoria), the 'deterrence trap' associated with fines is not avoided. However, such orders can still serve important retributive and restorative functions and, as such are valuable in sentencing corporate environmental offenders.

In NSW, the LEC often requires an offender to actually carry out the work associated with the project. Projects may relate directly to the environmental harm

[234] Above n 167.
[235] Above n 230, [23].
[236] See Table X above.
[237] EPA Victoria Annual Report, App, 10.
[238] EPA Victoria Annual Report, App, 8.

resulting from the offence, although it is questionable why, with the availability of clean-up orders, ESOs would need to be used in this way. In the LEC in NSW, for example, Simplot Australia Pty Ltd was found guilty of polluting waters with food waste.[239] The company was required to comply with two ESOs. At a cost of $20,000, the company was required to undertake restoration work of the river environment in the vicinity of the location where the initial offence took place. Under the second order, the company had to undertake an assessment of ways of reducing greenhouse gas emissions from its coal-fired plant to a level which was at or below 108 per cent of the estimated levels emitted from the premises in 1990.

vi The Development of Environmental Service Orders in England and Wales

At the time of writing, courts in England and Wales cannot require corporate or individual polluters to undertake a community improvement project. There is, however, clear support for the introduction of some kind of ESO. Macrory, in his final report on Making Sanctions Effective, supports corporate community service orders and emphasises the fact that by using these orders, the business community can take responsibility for its actions and restore any harm caused by the regulatory infraction. He advocates, more specifically, an order which would require:

> [A] business to complete an appropriate community improvement project within a specified period and for a specified value related to the underlying harm or benefit that has been caused or obtained by the offender.[240]

The Environment Agency in England and Wales indicated its support for what have been termed 'community project orders' which would be applicable to both corporate and individual offenders.[241] So in addition to the approach advocated in the Macrory review, the Agency has highlighted how the order would be useful in dealing with individual non-corporate offenders who may not be able to pay a high fine:

> Not infrequently the Agency encounters such individuals who may well have caused considerable environmental and amenity damage through fly-tipping for instance but who are not in a position financially to pay an appropriate fine. The ability of a Court to make [a community project order] would help to rectify the problem.[242]

The Agency has also recognised the other benefits flowing from this type of sentencing power. Environmental service orders indirectly ring-fence monetary penalties under the criminal law. In countries like England where fines are paid into central government funds, this is a particularly important advantage. The orders may also avoid the capacity problem associated with monetary fines whereby the value of the fine is limited by the wealth of the offender: an environmental service

[239] Above n 230.
[240] Above n 139 (Macrory Report), para 4.57.
[241] House of Commons Environmental Audit Committee, *Environmental Crime and the Courts* HC (6th Report of Session 2003–04) 126. Supplementary Memorandum from the Environment Agency.
[242] *Ibid* [7].

order could be useful where an offender does not have the financial capacity to pay a cash fine, but could absorb the costs of undertaking project work. The main objection to these orders is that, in contrast with a straightforward monetary fine, there will clearly be administrative costs associated with monitoring the progress of the project, expenses which, in the usual course of events, would be borne by the regulator. But on balance, the benefits would appear to outweigh the disadvantages and one can see little objection to giving the English courts a similar power.

E Rehabilitative Sanctions

In theory, fines, adverse publicity orders and community sentences can lead to organisational reform. However, in practice, such reform may not occur. To address this, creative sentencing options could be designed to target organisational and management deficiencies by, for example, requiring the company to address corporate operating and incident response procedures. Such options would clearly serve a rehabilitative function although, as will be discussed below, they also display important deterrence-based characteristics.

Federal and provincial legislation in Canada and Australia provides for a range of different types of rehabilitative sanctions. In some jurisdictions, the court can require offenders to perform an environmental audit which will not only improve management practices within the company but will also make the enforcing authority aware of any possible future violations. So, in many respects, the primary purpose of such orders is not to punish or deter the offender, but to prevent the recurrence of an offence. However, as the offender will have to pay for any audit carried out at the request of the court, this raises the value of the expected penalty in the economic deterrence framework. Mandatory audit orders also respond to one of the key criticisms of fines, that is their inability to require a review of internal company procedures. By requiring companies to identify defective procedures and furnish the regulator with specific information about the process, the courts can indirectly minimise the risk of future environmental incidents.

In Canada, explicit provision for an environmental audit order is made under section 291(1)(f) of CEPA. In Ontario, such an order can be required under general powers contained in section 72(1)(b) of the Provincial Offences Act. Such orders are extremely rare and I can find no recent examples of orders being issued under either of these two provisions. However, in 1992 a company was ordered to allocate $5,000 to hire an independent consultant to perform an environmental audit and to train employees.[243] In an even earlier case, the Ontario Ministry of the Environment was given access to the defendant company records, vehicles and storage equipment for the purposes of an environmental audit at the defendant's expense.[244]

[243] *R v Bio-Med Waste Disposal System Ltd and Jason Hwang*, unreported, Ontario Court Provincial Division, 10 March 1992 .

[244] *R v OC Liquid Waste Haulers of Ontario Ltd*, unreported, Ontario Court Provincial Division, 17 May 1990.

In Australia, both the EPA 1970 (Victoria) and the PEOA 1997 (NSW) empower the courts to impose an environmental audit order, under which the offender must carry out a specified environmental audit of activities carried on by that person.[245] In short, the purpose of audit orders is to ensure that the defendant's operations are environmentally acceptable. An audit order will be most appropriate when 'the defendant's operations lack essential environment protection systems or that there are serious ongoing failures in those systems.'[246] However, notwithstanding the benefits inherent in requiring an environmental audit, examples of this order are, as in Canada, difficult to find. One rare example relates to the prosecution of Calleja Nominees Pty Ltd by the EPA Victoria for failure to comply with vehicle emission standards. The company was ordered to pay $4,400 to carry out an environmental audit of its fleet of vehicles.[247] The LEC in NSW will occasionally require an offender to carry out an environmental audit although, as in Victoria, examples of such orders are few and far between. Recently, the LEC ordered that an environmental audit be carried out by a company that was found guilty of emitting offensive odours from its site. The company was required to conduct an environmental audit of its premises, the results of which were to be submitted to inter alia the LEC, made available for inspection at the company's administrative offices, and published on an Australian website.[248] It is somewhat surprising that the judicial power to order an environmental audit is not used more frequently in NSW as it would seem to be a particularly useful sentencing tool in this jurisdiction. Why? Due to the delay between the commission of an offence and the criminal hearing, it may be more effective to invest an auditing power in the regulators rather than the courts. Chapter 8 will in fact discuss the auditing powers of regulators. However, in NSW in particular, the judicial power is far more extensive, thus making it a particularly useful tool. First, the courts can extend the audit to activities not directly connected with the offence in question. And second, the regulators' powers are limited to where provision is made for such action in a licence condition, where the offence relates to operating without a licence or where an offender is exempt from licensing requirements but breaches general pollution laws.[249]

Corporate rehabilitation could also be the primary goal of orders targeted at improving environmental management, training and education. This may be achieved by requiring the defendant company, for example, to install an accredited environmental management system.[250] In 1998 for example, Prototype Circuits Inc. was prosecuted by CEPA for unlawfully importing and exporting hazardous waste. The company was ordered to implement an environmental management

[245] In Victoria see EPA 1970 s 67AC(2)(d) and in NSW PEOA 1997 s 250(1)(d).

[246] Above n 167.

[247] Note that under the legislation, audits can relate to any premises and activities carried on by the offender and are not confined to an audit of the conditions surrounding the offence itself.

[248] *Environmental Protection Authority v Shoalhaven Starches Pty Ltd* [2005] NSWLEC 684, [53], [55] and Annexure B. Cited in Preston above n 18, 163.

[249] For a discussion of administrative environmental audit powers in NSW, see ch 8 s III C.

[250] See, eg, s 291(1)(e) CEPA (Canada)

system consistent with the ISO 14001 standard.[251] In NSW, the LEC can order an offender to attend a training course (or other course) specified by the court or to establish, for employees or contractors of the offender, a training course of the kind specified by the court.[252] At the time of writing, it seems that the LEC has not exercised its powers in this respect. In some states and provinces such as Alberta, Canada, rehabilitative orders may be targeted not only at improving practices and procedures with the offending company but may also be of benefit to other operators. For example, in a prosecution brought by Alberta Environment, Buffalo Airways Ltd was fined $10,000 and given a creative sentencing order of $40,000 for unlawfully disposing of hazardous waste. The order required the company, in conjunction with the Buffalo School of Aviation, to develop a training programme for individuals to learn the proper process of aircraft paint-stripping.[253]

As with many of the other creative sentencing options discussed so far, rehabilitative sanctions of this type are not currently available to courts in England and Wales. However, there has been much support for the introduction of corporate sanctions that have the potential to rehabilitate offenders. Macrory recommends the introduction of corporate rehabilitation orders, the use of which:

> [G]oes beyond what a fine can achieve . . . by identifying tangible steps that a company must take and binding that company to their implementation. In taking steps to solve a company's compliance problems (as opposed to simply fining) there is also a greater chance that the individuals responsible will be identified and be held to account for their offences.[254]

These corporate rehabilitation orders would last for a specified period of time, during which the company would have to undertake specific actions and activities including, for example, the training of personnel in environmental issues, the adoption and implementation of environmental action plans and the remediation of environmental harm. Mandatory audits could form part of such an order. According to Macrory, the regulatory authority could maintain a list of accredited agents who could carry out the order, the costs of which would be borne by the offending company.[255] Macrory also advocates that community projects be incorporated within a corporate rehabilitation order.

F Conclusion

In this section I have considered a wide range of creative sentencing options available in Australia and Canada. All these tools clearly supplement and complement more traditional criminal penalties such as fines and custodial sentences, and give

[251] Environment Canada, *CEPA 1998 Legal Activities Report: Prosecutions Carried out from January 1 1988 to March 31 1999.*

[252] See PEOA 1997 s 250(1)(f), (g).

[253] Alberta Environment, *Enforcement Actions April 1 to June 31 2004*, 9.

[254] Above n 139 (Macrory Report), para 4.56.

[255] *Ibid* para 4.63.

the courts a degree of flexibility in tailoring the sentence to the specifics of the offence and the offender. In all cases, the sanctions serve a deterrence value. By removing any monetary benefit accruing to the offender (or uility), profit orders, for example, can clearly be factored into the deterrence framework. Compliance with other creative sentencing orders will impose costs on the recipient; costs which can be factored into the value of the sanction (D) in the deterrence framework developed in Chapter 2. However, the extent to which these sanctions are actually used in practice varies considerably from one sanction to another, and jurisdictionally. The less likely it is that a sanction will be used, then the lower its deterrent effect. So, for example, environmental audit orders are rarely used and as a result have a limited deterrent value. It is, of course important to remember that many of these sanctions reflect other penological principles most notably rehabilitation, restoration and reparation. In the context of corporate crime, these can be viewed as subgoals of deterrence. Sentencing options in England and Wales are much less developed. Many of the creative sanctions discussed in this section are not available to the courts, although there is considerable support for a widening of judicial sentencing powers. In my view, this would not only strengthen the deterrent impact of pollution control enforcement, but would also potentially improve the environment, educate offenders and promote future compliance.

VII Civil Enforcement

This final section considers the availability and use of civil enforcement proceedings in pollution control regulation in Australia, Canada, and England and Wales. In other words, it examines how and when a public body can take civil action for breach of a regulatory provision. In the pollution control context, it is very rare for a regulator to instigate civil proceedings against an individual or company. Their deterrent impact is therefore extremely limited and for this reason, civil enforcement is only discussed briefly.

Most jurisdictions enable regulatory authorities (and other parties as prescribed by legislation) to take civil action for an injunction to stop or prevent a violation of the legislation.[256] So, for example, under Division 14 of the EPBCA (Australia), the Minister can apply to the court for an injunction. That injunction can be prohibitory, in that it can restrain the recipient from engaging in conduct that

[256] In Australia see, eg, EPBCA (federal Australia) Division 14, PEOA 1997 (NSW) ss 252, 253, and EPA (Victoria) s 64A . In Canada see, eg, CEPA s 311. And in the UK see, eg, Pollution Prevention and Control Regulations 2000 SI 2000/1973 reg 33, and EPA 1990 s 38(1)–(13), which allows the Environment Agency to take enforcement proceedings in the High Court to obtain an injunction which forces a person whose licence has been suspended to carry out the required work. In addition, injunctive relief is often available to those persons whose interests are affected, who have suffered loss or damage to property or who have incurred expenses in attempting to mitigate or even prevent such loss. In some jurisdictions, interested or concerned citizens who have in fact suffered no particular damage may also be able to bring civil proceedings, either by right or with leave of the court.

constitutes a contravention of the Act. Alternatively, a mandatory injunction may be issued where a person is refusing or failing to do an act, the refusal or failure of which constitutes an offence. The injunction will require the recipient to carry out the specific act. From a theoretical perspective, civil injunctions clearly have advantages. First, and in stark contrast with criminal penalties, injunctions are proactive and are therefore particularly useful in preventing anticipated harm. And second, the courts are able to award an injunction at relatively short notice and where the terms of such an order are breached, penalties (including imprisonment for contempt of court) are severe.

They do, however, have a number of limitations. Although the right to apply for an injunction is not conditional on bringing prosecution proceedings against the 'offender', it is an equitable remedy and is available at the discretion of the court. For example, in exercising its discretionary power, both the UK and Australian courts apply the 'balance of convenience' test, balancing the advantages of an award to the claimant against the disadvantages to the respondent.[257] In cases where any environmental damage would be irreversible, the Australian courts seem more willing than their UK counterparts to give greater weight to protection of the public interest as opposed to hardship suffered by the respondent.[258] However, the author's discussions with EPA officers and environmental law academics indicated that regulators will rarely exercise their right to apply for an injunction. The primary reason for this is that some administrative enforcement powers such as the serving of notices or the suspension or revocation of a licence perform a very similar role to the injunction. In the UK, the courts have stressed that regulators must exhaust all criminal remedies before seeking an injunction.[259] Injunctions are therefore viewed as impractical and unattractive by regulators.

In Australia, the EPBCA 1999 includes civil penalty provisions. This is an enforcement strategy which is common in other areas of law (such as trade practices) but has not been used in pollution control regulation and, indeed, environmental law more generally in Australia, Canada or the UK. These civil penalty provisions include proscriptions against taking, without approval, action that is likely to have a significant impact on matters of national environmental significance and breaching the conditions of an approval granted under the Act.[260] The provisions can be enforced either by applying to the Federal Court for an injunction (discussed above) or an order that the wrongdoer pay a prescribed penalty.[261] In fixing the penalty, the court must take into account various factors such as the

[257] For further information on the 'balance of convenience' test in Australian environmental cases, see G Bates, *Environmental Law in Australia*, 5th edn (Australia, Butterworths LexisNexis, 2002) 170–72. Also, in some cases, the applicant may have to give an undertaking in damages whereby the applicant may have to compensate the respondent for any losses if the final decision goes against her.

[258] *Ibid* 170.

[259] *Tameside MBC v Smith Bros (Hyde) Ltd* [1996] Env LR 316.

[260] See EPBCA 1999 ss 12, 142.

[261] See EPBCA 1999 ss 481, 485 . The term 'wrongdoer' has been interpreted broadly. See further R Lyster, Z Lipman, N Franklin, G Wiffen and L Pearson, *Environmental and Planning Law in New South Wales* (Sydney, The Federation Press, 2007) 181.

nature and extent of the contravention and the circumstances in which the contravention took place.[262] The Minister may also accept an enforceable financial undertaking from a person who the Minister considers has breached a civil penalty provision.[263]

Pollution control legislation in Australia and Canada also empowers the courts to compensate regulatory authorities and members of the public who have suffered loss or damage to property as a result of the breach, although, rather than commence separate civil proceedings, it may be more cost effective to recover compensation when the offender is prosecuted.[264] Under the legislation, those affected may be able to recover reasonable costs and expenses incurred as a result of taking action to prevent or mitigate any loss.

VIII Conclusion

In this chapter, I have evaluated the effectiveness of both monetary and non-monetary criminal sanctions in deterring polluters, and considered, briefly, civil sanctions such as injunctions and civil monetary penalties. Under the deterrence framework, rational actors are induced to comply with the law if the benefits accruing to the offender from the offence are outweighed by the expected penalty (as determined by the sanction discounted by the probability of prosecution and conviction). This framework not only provides a valuable way of assessing enforcement policy and practice, but also enables us to compare the effectiveness of different criminal law sanctions in inducing compliance. Importantly, environmental enforcement strategies in Australia, Canada, and England and Wales display the key characteristics of a compliance-based approach to enforcement in which the use of formal enforcement tools, including criminal prosecution, is relatively rare when compared with the number of pollution incidents. Although regulators are normally successful in the prosecutions they bring, the infrequent use of this enforcement tool discounts any sanction that is imposed by the courts.[265]

This makes it all the more important for criminal sentencing to send a strong deterrent message. The Australian, Canadian and English/Welsh judicial approach to sentencing environmental offenders is broadly comparable in many ways. They take into account similar mitigating factors when determining an appropriate sentence, they rely predominantly on the use of criminal fines and they can utilise other 'traditional' penalties such as custodial sentences and community-based sanctions. The main difference, however, lies in the range of criminal sanctions

[262] EPBCA 1999 s 481.
[263] S 486DA.
[264] See PEOA 1997 (NSW) ss 246, 247. See generally s V B i above.
[265] See generally ch 5.

available to the courts. In Australia and Canada, provision is made in the legislation for both more 'severe' traditional penalties in the form of, for example, higher maximum fines, and a range of creative sentencing options such as adverse publicity orders, profit orders and environmental project orders, many of which are targeted at corporate offenders who cannot be imprisoned or subjected to other non-custodial alternatives. All these orders can be used in conjunction both with each other and with traditional penalties such as fines. Creative sentencing options for environmental crime are not yet available to the courts in England and Wales. There is, however, strong support for a broader, more flexible range of sentencing options, and it is likely that many options will be introduced in the foreseeable future.

From a deterrence perspective some of these alternative sentencing tools are little more than monetary penalties, and as such do not overcome many of the limitations of criminal fines. Sentencing options such as profit orders, restoration and prevention orders and orders requiring the offender to pay a sum of money to, for example, an environmental project or fund are still limited in the extent to which they deter environmental crime. They will be viewed by many corporations as nothing more than the cost of doing business, the 'deterrence' trap will not necessarily be avoided, and the full force of the penalty may be felt by innocent third parties such as consumers and shareholders as opposed to the responsible officers within the corporation. These alternative monetary penalties do however have a number of advantages over fines. Profit orders in particular, by removing the incentive on business to continue to fail to comply, can strengthen the deterrent effect of the criminal justice system. And restoration, prevention and environmental project orders ensure that any financial penalty is directed to environmental improvement of some kind, thereby reflecting a penological approach based more on restorative justice

There is also a host of non-monetary alternative sentencing tools such as environmental audit orders, adverse publicity orders, environmental project orders (where the offender is actually required to carry out a project) and corporate rehabilitation orders (as proposed by the Macrory Review). These options not only strengthen the deterrent impact of sentencing but also expand the variety of rehabilitative, restorative and retributive measures available against corporations and in doing so circumvent some of the major limitations of monetary fines. Many non-monetary creative sentencing orders avoid both the deterrence trap and spillover effects, and potentially underline the social unacceptability of corporate environmental offences. Unlike fines, they can reflect the non-financial motivations of the firm and possibly force a change in the behaviour of the offender. And as importantly, they give the courts flexibility sentencing offenders in a way that best reflects the offence and the offender.

This is not to suggest that fines or indeed other monetary penalties have no useful role to play in sentencing environmental offenders. Fines may well be an 'expedient and adequate solution' especially in dealing with less serious offences and it may be appropriate to combine one or more creative sentencing option with a monetary fine.[266] Furthermore, the use of creative sentencing is not free from difficulty. Non-financial sentences such as environmental project and audit orders may require a degree of regulatory and/or judicial monitoring which is not necessary in the case of a fine. This will obviously affect the cost-effectiveness of such sanctions. However, provided that a suitable statutory framework is drafted so as to maximize the effectiveness of creative sentencing, I believe that the courts should be able to design a sentencing package which will promote deterrence whilst at the same time reflecting other important goals of the criminal justice system. Regulatory, judicial or legislative guidance may also be appropriate in assisting the courts (and regulators) in applying the creative sentencing provisions.

Of course, criminal prosecution is just one of the enforcement tools available to regulators in dealing with non-compliance. The fact that it is used rarely in the environmental sphere may reflect the high costs (and risks) associated with using the criminal courts. Some commentators would argue that a much more effective way of deterring environmental crime is to utilise administrative enforcement mechanisms such as statutory notices and environmental penalties.

266 Above n 35, 249.

8

Administrative and Deterrence Sanctions

I Introduction

THE CRIMINAL JUSTICE process clearly plays an incredibly important role in deterring undesirable behaviour. Although pollution control regulators in Australia, Canada, and England and Wales use criminal prosecution as somewhat of a 'last resort', its mere presence towards the apex of the enforcement pyramid is arguably enough to move potentially recalcitrant firms towards compliance. And where a prosecution is initiated, the reputational consequences of conviction and the financial and non-financial sanctions available to the court send a strong deterrence signal. But equally important in the regulatory enforcement arsenal are formal administratively-imposed sanctions, named as such by virtue of the fact that they can be utilised without recourse to a judicial body. The precise nature and extent of these administrative powers varies considerably from one state to another, but broadly speaking, they can be divided into two categories. First, there are monetary administrative penalties which impose a direct financial sanction on offenders. Second, there are non-financial sanctions such as administrative notices or orders, licence suspension/revocation and enforceable undertakings which require the offender to carry out certain action. This chapter examines these administrative enforcement powers using, as before, the deterrence framework in Chapter 2.

Before considering these administrative tools, let us begin by reminding ourselves of the deterrence model. Becker's deterrence hypothesis was traditionally advanced as a model which predicted when criminal prosecution and criminal penalties would deter unlawful behaviour. But the model can also be applied to other types of sanctions, not just those consequent to a criminal conviction. This provides an even more useful evaluative tool for our purposes, as in the majority of cases, an enforcing authority will not resort to criminal prosecution and even if it does, will progress through several stages of enforcement before reaching the criminal courts.[1] So, in order to reflect the realities of the pollution control enforcement process, it is desirable to reconsider Becker's approach by factoring in the likelihood and cost of being caught (such as the costs incurred in complying with an

[1] See generally ch 3 for a discussion of enforcement strategies, and ch 5, which examines the enforcement approaches of selected environmental regulators in Australia, Canada and England/Wales.

administrative sanction), alongside any expected penalty that is imposed by a court. The deterrence hypothesis can therefore be rewritten as such:

$$B < gH + eF + cD$$

where gH represents the probability and associated costs of detection by an enforcement agency (including the hassle costs associated with detection), eF the probability and associated costs of administrative enforcement action (such as the imposition of a monetary administrative penalty or an enforcement notice) and cD the probability of formal court action (both civil and criminal) and its associated costs (including a court-based sanction).[2] This chapter is concerned, then, with the deterrent impact of eF.

Administrative enforcement mechanisms have a number of distinct advantages over their criminal counterparts. The first and probably the most obvious is that the administrative process consumes less time and therefore fewer resources than the criminal process. Of course, serious cases should be pursued in the criminal courts. But where prosecution would be considered heavy handed, an enforcing authority could utilise, relatively quickly and cheaply, a number of financial and non-financial administrative sanctions. Administrative mechanisms therefore represent a cost-effective enforcement option. The cost of implementing a particular sanction is also relevant from a deterrence perspective for as administrative sanctions are far more likely to be used than criminal prosecution, this raises the value of e in the deterrence framework. The relative speed of the administrative process is, in itself, important for a number of other reasons. Regulation will be most effective where there is co-operation between the regulatory agency and the regulated community. As noted by Brown and Rankin, the official who recommends a sanction and the accused actor are 'cast in the role of adversaries while enforcement proceedings are pending'.[3] This may stem the free flow of information needed to work out solutions to regulatory problems. In addition, administrative sanctions can, unlike criminal prosecution, respond quickly to emergency situations. In such circumstances, administrative notices and orders can be extremely effective in reducing the risk of damage and, where appropriate, remediating harm that has already occurred. The second strength of administrative sanctions is that they are imposed, not by the courts, but by specialist regulatory officials who are more familiar with the legislative framework, the scientific and technological processes relevant to the regulated entity's operations and the personal circumstances of the offender.

[2] A Ogus and C Abbot, 'Sanctions for Pollution: Do We Have the Right Regime?' (2002) 14 *Journal of Environmental Law* 283, 290.

[3] R Brown and M Rankin, 'Persuasion, Penalties and Prosecution: Administrative v Criminal Sanctions' 346, in ML Friedland (ed), *Securing Compliance: Seven Case Studies* (Toronto, University of Toronto Press, 1990).

Pollution control regulation in Australia, Canada, and England and Wales makes provision for a range of administrative enforcement sanctions. The Environmental Protection Authority (EPA) Victoria, the Department of Environment and Climate Change (DECC) in New South Wales and the Department of Environment, Water, Heritage and the Arts (DEWHA) at federal level all have the power to serve a range of administrative sanctions with enforcement statistics pointing strongly to the fact that these powers are used far more frequently than criminal prosecution.[4] One of the most innovative enforcement tools used in Australia is the enforceable undertaking.[5] Enforceable undertakings are written promises between the regulator and the offender, in which the offender commits to carry out a range of measures that are directed at promoting future compliance, remediating existing harm and compensating victims of the offence. Administrative mechanisms are also available in Canadian pollution control regulation. Like current practice in Australia, they are also used more commonly than prosecution[6] although, as noted by Benedickson, except where there is an emergency situation, 'administrative orders are generally resorted to after discussion and consultation have failed to resolve an environmental situation.'[7] In contrast, the Environment Agency in England and Wales seems to prosecute more frequently than its counterparts in Australia and Canada, and is less likely to use administrative sanctions.[8] One reason for this could be that the range of administrative sanctions available to the Agency is far more limited. Administrative notices and licence suspension and revocation have been important components of the enforcement toolbox for some time. But until recently, monetary administrative penalties and enforceable undertakings were notably absent from the law enforcement framework. However, following several government reviews of regulation and subsequent calls for the introduction of a range of new administrative sanctions,[9] Part III of the Regulatory Enforcement and Sanctions Bill (now the Regulatory Enforcement and Sanctions Act 2008) introduces, amongst other things, monetary administrative penalties and enforcement undertakings. This aspect of the Bill was generally warmly received by both Houses of Parliament, with Lord Jones of Birmingham making the following observation:

> [Part 3] creates a framework within which regulators can acquire access to a suite of civil [administrative] sanctioning powers while not hurting their current criminal sanction capability. Not all the powers will be necessary in all cases; sometimes comparable powers already exist, and sometimes they will not be appropriate in a particular context. The Bill therefore allows Ministers, by order, to select particular options from the range of

[4] See further ch 5 s II B and C.

[5] Enforceable undertakings are discussed in detail in s III D below.

[6] See further ch 5 s III B and C.

[7] J Benidickson, *Environmental Law* (Toronto, Irwin Law, 2002) 127.

[8] See further ch 5 s IV.

[9] See, eg, Hampton Review, *Reducing Administrative Burdens: Effective Inspection and Enforcement* (March 2005); Macrory Review, Better Regulation Executive, *Regulatory Justice: Sanctioning in a post-Hampton World* (Consultation Document, May 2006) and *Regulatory Justice: Making Sanctions Effective—Final Report* (November 2006); DEFRA, *Review of Enforcement in Environmental Regulation, Report of Conclusions* (October 2006).

> new civil [administrative] sanctions, depending on the particular regulatory regime, and the regulator, whoever it may be, will be obliged to use those sanctions when they could usefully complement their criminal sanctions. This is flexible. This is new. This is different. This is a way of making the punishment fit the crime.[10]

The chapter begins in Section II by examining monetary administrative penalties. It then considers non-financial administrative sanctions such as administrative notices and enforceable undertakings in Section III.

II Monetary Administrative Penalties

A Introduction

In many jurisdictions, environmental regulators have the power to require the payment of a monetary administrative penalty (MAP) which, as is indicated by the use of the word 'administrative', can be imposed without recourse to the criminal or indeed civil courts. The precise characteristics of MAPs and the process by which they can be levied varies according to the enabling Act. In minor, high-volume instances of non-compliance, the size of the penalty will often be fixed at a relatively low level. These fixed monetary administrative penalties (often referred to as tickets or on-the-spot fines) will normally be imposed by individual inspectors or field staff. In more serious cases, the value of the penalty may be determined at the discretion of the regulator, and will often be variable so as to reflect the mitigating and aggravating circumstances of the particular case. Here, the decision as to the penalty may be taken independently from those regulatory officials working on the ground. This is conducive to consistent, transparent and accountable decision-making, and will serve to ensure that the relationship between regulator and the regulated entity is not compromised. The procedures by which a penalty can be challenged will also vary. In the United States, for example, regulators can levy a financial penalty without involving the courts, with the recipient of the penalty having a right to appeal to a second administrative agency. In contrast, in Australia, they have adopted what Macrory terms a 'hybrid' model of infringement notices or tickets[11], described by the Australian Law Reform Commission as:

> An administrative device to dispose of a matter that involves a criminal or non-criminal breach. When such a breach is committed, the relevant agency may prosecute or take civil penalty proceedings, or may issue an infringement notice offering the offending party the chance to discharge or expiate the breach through payment of a specified amount.[12]

[10] Second Reading of the Regulatory Enforcement and Sanctions Bill, House of Lords (28 November 2007 Col 1243). Text in square brackets added by the author.

[11] Macrory Consultation Document (above n 9) 53–4 refers to these as 'applied monetary penalties'.

[12] Australian Law Reform Commission Report 95, *Principled Regulation: Federal Civil and Administrative Penalties in Australia* (2002) 78.

Despite the fact that there are important differences in the characteristics of these MAPs, one feature that they all have in common is that they can be applied directly by the regulatory authority. Being an interim step between the formal and costly action of criminal prosecution and more informal means of advice and persuasion, MAPs are a useful addition to the enforcement toolbox. In section B below, I examine how such a penalty device fits within the deterrence framework and consider more generally the strengths and weaknesses of MAPs. I then provide an overview of MAP provisions in Australia, Canada, and England and Wales.

B Strengths and Weaknesses

As discussed previously, a primary advantage of administrative enforcement tools generally is that they can be utilised at a lower cost than criminal prosecution. As the costs of criminal prosecution are high, regulators are only able to pursue a small number of cases through the criminal courts. Furthermore, criminal prosecution may not be appropriate or in the public interest. In contrast, MAPs provide 'relief from the high number of prosecutions that would otherwise have to be conducted to enforce the law relating to summary offences'.[13] In other words, the legal cost associated with sanctioning a firm or individual for non-compliance through the levying of a MAP is, compared with criminal prosecution, much lower for both government and industry. Bagaric describes this as the 'cornerstone' of an argument that supports an expansion of the range of matters which can dealt with by way of an on-the-spot fine in Australia.[14] The fact that the legal costs of levying a MAP are less than those associated with prosecution significantly impacts on the deterrence framework. Presuming that non-compliance is detected in the first place, the availability of MAPs may raise the value of e (the likelihood of imposition) in the amended deterrence model. The actual value of a MAP (F in the deterrence framework) may, of course, be lower than any monetary (or other) sanction determined by a court consequent to criminal prosecution, especially where the MAP is fixed as opposed to variable. But this must be weighed against the fact that the lower costs associated with levying a MAP (compared with criminal prosecution) mean that there will be a higher probability of imposition. In other words, as more recalcitrant firms can be sanctioned and brought into compliance, the deterrent effect of a MAP system will normally be greater than the deterrent effect of prosecution. MAPs may also improve deterrence by removing or reducing the profit-making component of regulatory non-compliance. This is important under the deterrence framework as any profits drawn from non-compliance are a crucial part of the cost-benefit equation. Also, in the case of variable MAPs in particular, the level of penalty applied is determined by the regulator, a competent, expert

[13] R Fox, 'Infringement Notices: Time for Reform' (1995) 50 *Trends and Issues in Criminal Justice* 1, 4.

[14] M Bagaric, 'Instant Justice? The Desirability of Expanding the Range of Criminal Offences Dealt with on the Spot' (1998) 24 *Monash Uni Law Review* 231, 250.

agency who is arguably in a better position than the courts to levy a penalty which best reflects the circumstances of non-compliance.

Brown and Rankin, in examining criminal and administrative penalties as used by the Waste Management Branch and the Workers' Compensation Board in British Columbia, cite a number of other reasons why MAPs may be more effective than criminal prosecution. First, they argue that MAPs are more effective when operating within a risk-based, as opposed to a harm-based, regulatory system. Many environmental infractions occur not because harm has actually occurred, but because there is a risk of harm and regulators, who focus much of their inspection efforts on identifying hazards before harm occurs, arguably view risk more seriously than prosecutors and judges.[15] Second, Brown and Rankin point to the fact that regulators may be more likely to use MAPs because they are less stigmatising than a criminal prosecution.[16] This reluctance to prosecute can be linked to the real versus regulatory crime debate discussed in Chapter 6.

However, monetary administrative penalties are not a panacea; they suffer from many of the problems already identified in relation to fines imposed by the criminal courts.[17] Most importantly, the maximum fine in any one case is constrained by the wealth of the recipient (this is known as the 'deterrence trap')[18] and unless non-compliance is caused by a profit calculation, fines do not address the issue of future compliant behaviour. More generally, it could be argued that there is a risk of enforcement officials failing to consider the individual circumstances of each case, potentially leading to injustice. This has a particular bearing on the application of a variable MAP scheme. Such risks can, however, be minimised by ensuring that the system (and regulatory officials) operate within published guidelines. Doing so would also counter concerns about the damaging impact that a MAP system can have on the relationship between the regulated and the regulator. A second concern is that a MAP system will allow firms to buy their way out of compliance. But providing that 'there is transparency in the system, the firm's non-compliance will still be a matter for the public record and . . . industry will not be doing a deal with regulators behind closed doors.'[19] One final issue relates to whether or not the relative ease of serving a MAP will lead regulators to target legitimate business as opposed to 'rogue' operators. In response, Macrory points to a number of the advantages of MAPs discussed above, particularly their flexibility and the opportunity that the system affords to regulators to minimise the financial benefits of non-compliant activity, both of which give regulators the opportunity to deter rogue elements.[20]

All in all, MAPs should be seen as an addition to the regulatory enforcement toolbox. They do not, however, replace criminal prosecution. Whilst the stigma of

[15] Above n 3, 341.
[16] *Ibid* 341–2.
[17] This matter is discussed in detail in ch 7 s III.
[18] Discussed in ch 7 s III C.
[19] Macrory Consultation (above n 9), para 3.68.
[20] *Ibid* para 3.70.

criminal sanctions may cause regulators to refrain from prosecuting offenders, it also remains one of the most potent deterrents of the criminal justice system. Brown and Rankin conclude that:

> Regulators who can impose administrative penalties are much better equipped to tackle this compliance deficit than those who must resort to criminal prosecution. The administrative process responds to risk rather than to harm, does not unduly stigmatize offenders who are thought not to warrant moral opprobrium, applies a standards of absolute as opposed to strict liability in at least some cases, entails minimal operating costs, and imposes monetary penalties large enough to have a reasonable prospect of deterring offenders. Criminal prosecution, the most common sanction of last resort among Canadian regulatory agencies, scores poorly on all these counts.[21]

C Penalty Infringement Notices in Australia

Rather than take civil or criminal proceedings, Australian environmental regulators can, in certain circumstances, give the offending party the opportunity to discharge or compensate for the breach through the payment of a specified monetary amount.[22] The value of the fine appropriate for any given offence will normally be laid down in the implementing legislation and where the fine is paid, no criminal conviction is recorded.[23] These penalty infringement notices (PINs) therefore amount to fixed, as opposed to variable, monetary administrative penalties. In fact, variable MAPs are not provided for in pollution control regulation in NSW, Victoria or at federal level. PINs are usually available for relatively minor offences such as littering, and the fines imposed are correspondingly low. So, for example, infringement notices served under the EPA 1970 (Victoria) range from $110 to $5,500.[24] Saying that, the EPA 1970 (Victoria) for example does allow PINs to be used for certain pollution offences such as discharging waste at scheduled premises without authorisation.[25] In both Victoria and NSW, the penalty payable under a PIN depends on whether the offence has been committed by a corporation or individual, with higher penalties being imposed on companies.[26] This ensures that regardless of the size of business, a fixed MAP can still represent an effective deterrent. Rather than pay the fine, the recipient of an infringement notice can elect to have the matter dealt with by a local criminal court, where the potential penalty payable is considerably higher than the maximum available under a penalty notice scheme. Not surprisingly, very few individuals and companies choose this option

[21] Above n 3, 348.

[22] See, eg, EPA 1970 (Victoria) s 63B and PEOA 1997 ss 222–9 (NSW).

[23] For example, EPA 1970 (Victoria) Sch A . In NSW, penalty notices are only applicable to some Tier 2 offences and all Tier 3 offences. See further Protection of the Environment Operations (Penalty Notices) Regulations 2004.

[24] In Victoria, penalties levied by infringement notices are calculated on the basis of penalty units. At present, a penalty unit amounts to $113.42, although by virtue of the Monetary Units Act 2004, the unit value will be amended in line with the Consumer Price Index.

[25] See EPA 1970 (Victoria) s 27(1).

[26] See, eg, Protection of the Environment Operations (Penalty Notices) Regulations 2004.

as they are not willing to risk paying a much larger court fine and would rather avoid the adverse publicity associated with a formal criminal conviction. In NSW, out of the thousands of penalty notices served in 2003, only approximately 50 offenders elected to appear before the court.[27]

Penalty infringement notices, like administrative notices, are a commonly used enforcement tool although, as noted above, they are mainly used for minor offences. In NSW in 2006–07, a total of 2,067 PINs were served by DECC authorised officers, the overwhelming majority of which related to smoky and noisy vehicles, and littering from motor vehicles. The average value of PINs was approximately $333.[28] Statistics would, however, indicate that authorised local government officers are three times more likely to levy a PIN than DECC authorised officers. The number of infringement notices served in Victoria is even higher (although the statistics do not distinguish between notices served by EPA officials and local government officers). In 2006–07, a total of 20,991 notices were served, 20,804 of which relate to littering offences. Eighty notices were served for industry-related and waste transport offences.[29]

The model adopted in countries such as Australia (termed 'applied' MAPs), where MAP recipients can pay the fine or elect to have their case heard in the criminal courts, has its limitations, several of which affect their ability to deter undesirable behaviour. First, the model's close links to the criminal justice system mean that it can be perceived as a means of encouraging offenders to buy their way out of prosecution. Second, this system of applied MAPs does not have the same cost-saving advantages as fixed or variable MAPs. As the regulator has to have enough evidence to secure a criminal conviction before levying the applied MAP, the evidentiary burden on the regulator may not be lessened. Whilst the evidence needed to prove most if not all of these minor environmental offences will not necessarily be difficult to obtain, the mere fact that cases would revert to a criminal court if the recipient chooses not to pay the fine means that the sanction could not be applied at a significantly lower cost than prosecution. Furthermore, cases are being sent to criminal courts, where there is already an issue with low deterrence in sentencing. Finally, due to lack of judicial scrutiny, it is argued that the rigorous procedural protections usually afforded to criminal offences are absent. Whilst this is inevitably the case in the context of these applied penalty models, it provides a strong argument in favour of a cleanly stream-lined administrative process in which recipients of a MAP cannot elect to have their case heard in the criminal courts but can appeal to a body such as an administrative tribunal.

Despite these concerns, enforcement officers in both NSW and Victoria conveyed their support for the penalty notice scheme in the case of minor environmental offences. The Director of Legal Services, NSW commented that it is a very powerful tool, particularly for councils that tend to regulate the smaller

[27] Information provided during an interview with the Director of Legal services, EPA NSW, March 2004.

[28] DECC, *Annual Report 2006–07*, 36.

[29] EPA Victoria, *Annual Report 2006–07* App, 10.

companies.[30] They would therefore seem to provide a useful tool in tackling minor crime. Anecdotal evidence would seem to support this hypothesis. For example, when fixed MAPs were introduced into the cattle identification regime in Canada, evidence presented to the Macrory review suggested a 20–30 per cent rise in compliance rates.[31]

D Administrative Penalties in Canada

Pollution control legislation in Ontario and at federal level also empowers regulatory officers to serve a fixed monetary administrative penalty, otherwise known as a 'ticket'. Ticketable offences generally relate to planning, record keeping and reporting and where a ticketable offence continues for more than one day, a ticket can be issued for every subsequent day on which the offence continues. The recipient of the ticket can either pay the fine without making a formal court appearance, plead guilty but appear in court to request a lesser fine or submit a plea of not guilty, resulting in formal court proceedings. For example, officers within the Ministry of the Environment (Ontario) can issue a Provincial Offences Act notice.[32] Used in the case of relatively minor violations, a ticket initiates a prosecution but permits direct payment of a fine if the person responsible pleads guilty. Under CEPA, where there is a minimal threat to the environment, human health or life, tickets can be issued under the Contraventions Act and associated regulations.[33]

Until recently, variable monetary administrative penalties had no place in the enforcement of pollution control regulation in Ontario. Although section 182 of the EPA 1990 made provision for variable MAPs, the section was never brought into force. However, following a number of high profile industrial spills in Ontario[34], the government established, in April 2004, IPAT (the Industrial Pollution Action Team) with a mandate to 'examine the causes of industrial spills and dangerous air emissions and recommend to the government prevention measures for industry and others.'[35] This review renewed calls for inter alia an administrative penalty regime and led ultimately to Bill 133, otherwise known as The Spills Bill.[36] As was made clear by the Minister of the Environment and others, the purpose of the Bill was not to punish industry but to encourage compliance by providing an important addition to the armoury of enforcement tools available under the EPA 1990 at the time. An equally important justification for the bill was to ensure that the remediation of any environmental impact was financed by the polluter and not the community.

[30] Information provided during an interview with the Director of Legal services, EPA NSW, March 2004.

[31] Macrory Consultation (above n 9), para 3.35.

[32] Provincial Offences Act (Ontario) s 3.

[33] Contraventions Regulations (SOR/96–313).

[34] In 2004, 139 facilities (currently covered by the municipal industrial strategy for abatement) were responsible for nearly 98% by volume of all reported liquid industrial spills in the province.

[35] Industrial Pollution Action Team, *Final Report* 3.

[36] The Bill was introduced in the Legislative Assembly of Ontario on 27 October 2004.

The Bill, now the Environmental Enforcement Statute Law Amendment Act 2005 was passed on 9 June 2005. The Act amends the EPA 1990 and in addition to providing stricter sentencing provisions,[37] empowers regulatory officials to impose an 'environmental' penalty in the event of certain contraventions. Details are provided in the Environmental Penalties Regulations 2007.[38] In terms of application, the environmental penalty provisions focus on those facilities responsible for a significant proportion of industrial spills to land and water in the early 2000s. More specifically, they apply to 148 facilities in nine 'Municipal Industrial Strategy for Abatement' sectors, including the petroleum, iron and steel and chemicals sectors. They do, however, exclude smaller firms who can, cumulatively cause serious damage to water and land.[39] The provisions are also limited to land and water violations.

The details of the regime (including how environmental penalties will be calculated and what types of contraventions will be covered) are set out in the 2007 Regulations. According to the Act, the maximum daily penalty applied to an individual will be $20,000 and in the case of corporations, could be as high as $100,000 a day.[40] Liability is 'absolute'. Due diligence and mistake of fact are not defences and a penalty can be applied even where the person took all reasonable steps to prevent the contravention or even when, at the time of the contravention, the person had an honest and reasonable belief in a mistaken set of facts that, if true, would have rendered the contravention innocent.[41] In the event of non-payment of the penalty it may, for example, be enforced as if it were an order of the court or the recipient of the penalty may have her certificate of approval, licence or permit suspended.[42] According to section 182.1(9), the imposition of an environmental penalty does not prevent the individual or company from being prosecuted for the same offence.

The Act also ensures that all environmental penalties paid are 'ring fenced' and are deposited into a separate account in the Consolidated Revenue Fund.[43] The Minister of the Environment can then direct that the money be paid out for a number of purposes. For example, persons who have incurred costs or expenses or suffered losses as a result of a pollutant spill may be compensated or persons who undertake environmental remediation projects may be provided with financial assistance.[44] An agreement between the Director and the recipient of the penalty may lead to its cancellation or reduction.[45] The Act does allow the recipient of an environmental penalty to request a review of the penalty. Such a review must be

[37] See ch 7 s III B ii esp Table VII for details.
[38] Ontario Regulation 222/07.
[39] This was a matter of concern expressed, amongst others, by Mr Barrett in the Bill's third reading (Hansard Debates 7 June 2005, 7540).
[40] S 182.1(5).
[41] S 182.1(6). For a discussion of defences available in pollution control regulation see ch 6 s III B.
[42] S 182.1(11).
[43] S 182.4(1).
[44] S 182.4 (3).
[45] S 182.1(8).

made within 15 days of the date indicated on the notice. Appeals against an order are also permitted under the Act and regulations although the onus of proof is on the violator to disprove the elements of the contravention.[46] Both appeals and reviews are heard by the Environmental Review Tribunal. Finally, the Act provides for the issuing of an annual report which will document each environmental penalty including the name of the person against whom an order was made, the amount of a penalty and a description of the contravention. The report will include the impact of any agreement entered into between the Director and the offender on the obligation to pay the penalty or the amount of the penalty.

The 2007 Regulations are very prescriptive in determining how an environmental penalty is calculated. It goes so far as to even prescribe a formula. The starting point is to determine any monetary benefit accruing to the offender from the offence. Added to this must be an amount to reflect the gravity or seriousness of the offence, as discounted by (a) any preventative or mitigative measures to prevent the spill or violation (b) the presence of an environmental management system at the time of the offence, and (c) the entering into any settlement agreement which leads to investment in an environmental project. The regulations tabulate offences according to their seriousness, and specify the maximum percentage reduction which can be achieved as a result of mitigating factors. Where the offence continues for more than one day, the penalty amount will again be determined using a specific formula. For most major violations, the penalty amount is not capped although it is anticipated that the average penalty will be approximately $1,000 for administrative offences and $20,000 for unlawful spills and spill-related offences.[47]

Described as a 'very controversial piece of legislation' by the Minister of the Environment, the Bill went through several stages of consultation and three readings in the Legislative Assembly before it became law.[48] Its provisions received broad support from many including the Ministry of the Environment and NGOs such as the Canadian Environmental Law Association (CELA) and the Sierra Legal Defence Fund. In the third and final reading of the Bill, the Minister of the Environment, Leona Dombrowsky, declared that its provisions would 'go a long way to encourage compliance in the industrial community and also to provide communities that are affected by these events with resources to address the emergencies when they occur.'[49] The Sierra Legal Defence Fund, Environmental Defence and CELA also applauded the introduction of the Bill, saying that 'polluters who want to stop paying will have to stop polluting.'[50] However, when the Bill was posted for comment under the Environmental Bill of Rights process, there were 164 responses, 157 of which expressed significant concerns about various

[46] S 140.

[47] Ministry of Environment (Ontario) News Bulletin, 'Environmental Penalties Regulations Help to Protect Ontario's Water Resources' (June 2007).

[48] Hansard HC col 1610 (2 June 2005).

[49] Hansard HC col 1530 (7 June 2005).

[50] Sierra Legal Defence Fund Media Release, 9 June 2005.

parts of the bill. Some NGOs and MPs argued that the Bill did not go far enough in tackling the pollutant spillage problem. In response to the noticeable absence of any proactive measures to deal with spills, one MP in describing environmental penalties as the 'spill and get a bill' approach, stated that the Bill was 'limited in its effectiveness as a virtual standalone method of deterrence.'[51] Many of the Industrial Pollution Action Team report's recommendations on pollution prevention and environmental management were in fact ignored.

Industry groupings (unsurprisingly) intensively lobbied against the introduction of environmental penalties, with some of their concerns also being voiced by MPs in the Legislative Assembly. Bodies such as the Ontario Chamber of Commerce were extremely concerned about the removal of due process (in that a person or corporate entity subjected to an environmental penalty has no recourse to the courts as a means of defending themselves) and due diligence as a defence, arguing that the Bill would 'threaten the viability of all existing business in Ontario, and furthermore harm any future possibilities for economic growth and investment.'[52] The availability (or not) of a due diligence defence in the context of administrative enforcement is a somewhat controversial issue, not only in relation to environmental regulation but in the context of other areas of law. Judge Bowman, in hearing a case involving administrative monetary penalties under the Excise Tax Act, made his position quite clear:

> That a person should be susceptible of being penalized administratively by a public servant without any possibility of exculpating himself by demonstrating due diligence is not only extraordinary. It is abhorrent. It is not less abhorrent because it is mechanically and routinely imposed by anonymous revenue officials and therefore qualifies for the essentially meaningless rubric 'administrative' rather than 'criminal'. A punishment is a punishment. Neither its nature nor its effect is tempered by the use of palliative modifiers.[53]

Despite such strongly voiced concerns, the environmental penalty provisions still impose 'absolute' liability on offenders, although the government has gone some way to acceding to demands from industry (and others) by allowing the offender's actions to prevent, minimise or expedite a cleanup to be taken into account when a penalty is considered. Placing the burden of proving remediation efforts on the polluter was also a matter of some concern. In defending this provision, the Minister of the Environment stated that it was the next logical step in dealing with a spill:

> If there is a spill at a company plant, then the company is in the best position to ensure that the spill is contained quickly, so that it does not violate environmental protection

[51] Hansard Debates, col 1640 (2 June 2005).

[52] See: <www.occ.on.ca/2policysubmissions/Bill%20133%20OCC%20Committee%20Submission.pdf> accessed.

[53] *Pillar Oilfield Projects Ltd. v R* [1993] 2 GTC 1005, 1009. See, more recently, *Consolidated Canadian Contractors Inc. v R* (1998) 231 NR 92, where Robertson JA accepted that, in certain circumstances, absolute liability may legitimately be attached to administrative penalties but that there was a rebuttable presumption that Parliament did not 'intend' to impose absolute liability.

laws. The company can put in place the appropriate monitoring and contingency responses to ensure its discharges remain within legal limits. If the company wants to challenge a penalty that the government has imposed for a spill, then the company is best positioned to demonstrate why its spill did not break the law. We contend that it is good public policy to emphasize spill prevention and expedite the cleanup of a spill, to get it done and paid for as quickly as possible.[54]

The government refused to back down on this provision and it was incorporated into the final draft of the Act. There was also some concern about the fact that any settlements agreed to between the Director and the recipient of the penalty (which contribute towards a reduction in the penalty) would fall foul of due process norms such as transparency and accountability.

In response to stakeholder consultation, the Bill actually underwent a series of substantial amendments which, according to many, substantially watered down its potential effectiveness in deterring unlawful spills. In total, 14 of the ideas brought forward during consultation and discussion were accommodated in the final Act. For example during the consultation process the business community expressed some concern in giving a provincial officer the power to levy a penalty.[55] The government responded to such concerns by amending the Bill and in its final format, only a Director of the Ministry of the Environment can impose an environmental penalty. In addition, the environmental penalty provisions do not apply to company directors and employees, thereby removing one of the most important deterrents: the ability to ensure that people responsible are held accountable.[56]

E Monetary Administrative Penalties in England and Wales

In England and Wales, fixed monetary administrative penalties (or 'fixed penalty notices') are available for a range of minor environmental crimes, such as littering and very small-scale fly-tipping, the money from which can only be used to help meet the cost of certain specified functions as provided for by legislation.[57] As is the case in Canada and Australia, where an offence has been committed (for which a fixed MAP is available), the offender is given the opportunity to avoid prosecution by paying the penalty. Legislation specifies who can serve a fixed MAP. Offences such as fly-posting, littering and abandoning a vehicle are all dealt with by local authorities. The penalty amount can be set at a local level, (within specific limits) although the legislation prescribes a default fine amount, normally £75. There are two offences for which Environment Agency officers are empowered to award a fixed MAP, namely failure to furnish a waste transfer note on request, and

[54] Leona Dombrowsky speaking before the Standing Committee on the Legislative Assembly (Hansard Debates 12 May 2005).

[55] Statement taken from evidence given by Perrin Beatty of the Coalition for a Sustainable Environment to the Standing Committee on the Legislative Assembly (12 May 2005).

[56] See ch 5 s IV A for a discussion of the merits of directorial liability.

[57] See, eg, Environmental Protection Act 1990 s 88(1) (littering) and Refuse Disposal (Amenity) Act 1978 s 2A(1) (abandoning a vehicle).

failure to produce a waste carrier's licence. Both can be dealt with by way of a £300 penalty.[58] Unlike Australian fixed MAP provisions, the amount payable by way of a fixed MAP does not vary depending on whether the offender is an individual or company. By not recognising the diversity of offenders' ability to pay, Macrory argues that the current fixed MAP provisions do not necessarily represent an effective deterrent. He therefore proposes that fixed MAPs should reflect factors such as the size of the firm (taking into account the number of employees or annual turnover).[59] In order to ensure that fixed MAPs are used only in appropriate cases, detailed guidance is provided by DEFRA.[60] The guidance emphasises the need to ensure that fixed MAPs are only used where there is adequate evidence to support a prosecution if a notice is unpaid, and that MAPs are not appropriate for repeat offenders or for more serious cases of, for example, fly-tipping where the offender would normally be prosecuted.

In recent years, there have been strong calls for giving environmental regulators the power to impose variable MAPs on environmental offenders. Both the Macrory and DEFRA reviews support the use of variable MAPs. Macrory proposes three variable MAP models, all of which reflect different enforcement and appeal routes. Table XI below, taken from the Macrory consultation document, broadly outlines the three models:

Table XI—Macrory Models of Administrative Penalties[61]

Model 1: Applied Monetary Penalties	Model 2: Fixed and Variable MAPs with appeals to the courts	Model 3: Fixed and Variable MAPs with appeals to an independent regulatory tribunal
• Regulator gathers evidence as if going to pursue criminal prosecution • The business has the option of paying the MAP or to let the regulator proceed with the case in court	• Regulators have access to fixed and variable MAPs • If the business does not agree with the penalty or merits of the case, it can appeal to the criminal courts	• Regulators have access to fixed and variable MAPs • If the business does not agree with the penalty or merits of the case, it can appeal to a specialist tribunal

Both Models 1 and 2 involve the criminal courts. Under Model 1 (which is similar to fixed MAP systems in Australia, Canada, and England and Wales) on refusal

[58] See Environmental Protection Act 1990 s 34A(2) and Control of Pollution (Amendment) Act 1989 s 5B(2) respectively.

[59] Macrory Final Report (above n 9), para 3.30.

[60] See eg DEFRA, *Fixed Penalty Notices: Guidance on the Fixed Penalty Notice Provisions of the Environmental Protection Act 1990, the Clean Neighbourhoods and Environment Act 2005 and other legislation* (2006) and DEFRA, *Local Environmental Enforcement—Guidance on the Use of Fixed Penalty Notices* (2007).

[61] Taken from Macrory Consultation (above n 9) Table 3.1, 55

to pay the penalty, the regulator will proceed with a criminal prosecution. As discussed in Section A above, this process reduces the deterrent effect of fixed MAPs. By involving the criminal justice process, fixed MAPs will not be as cost-effective as would otherwise be the case. Model 2 advocates the use of both fixed and variable MAPs, but if a firm does not agree with the application of the penalty, it would appeal to the criminal courts. This type of arrangement is not, according to Macrory, 'the optimal choice for reform because of the lack of separation between administrative sanctions and criminal sanctions'.[62] Involving the criminal courts at the appeal stage would also lower the deterrent impact of both fixed and variable MAPs and might overburden the already strained criminal justice system. Macrory favours Model 3, under which both fixed and variable MAPs can be appealed not to the criminal courts, but to an independent, specialist regulatory tribunal. This model, which is very similar to that introduced in Ontario, Canada, could, he argues, deliver better outcomes to both industry and regulators. It would ensure that appeal cases be heard by those with specialist expertise in environmental regulation and would remove the workload from (in England and Wales) magistrates' courts, which deal primarily with 'mainstream' crime. Also, by ensuring that MAPs would not be tied into the criminal justice system, it could remove the financial gains of non-compliance without the imposition of a criminal conviction. The majority of responses to this aspect of the Macrory review also came out in favour of Model 3.

Any system of variable MAPs would obviously have to be underpinned by a number of safeguards to ensure the system is fair, proportionate and transparent.[63] In the case of variable MAPs, it is crucial that in order to maintain transparency, the regulator (or other appropriate body) publish the methodology used in determining the appropriate penalty amount. A good example of this would be the detailed guidance published by the Ontario EPA, which stipulates the criteria used in calculating penalty amounts under the amended EPA 1990.[64] The exact methodology to be used would depend on the regulatory regime. Key factors would include the nature of the breach, the economic gain generated by non-compliance and the extent of potential harm, with the final penalty being determined by taking into account certain other aggravating and mitigating factors. It is also important to ensure that fixed and variable MAPs are imposed by the most appropriate person. In the case of fixed MAPs, the relatively low penalties involved means that decisions can legitimately be made by individual regulatory officers. However, it would be more appropriate for variable MAPs to be determined not by field staff, but by personnel which operate independently from staff on the

[62] Macrory Final Report (above n 9), para 3.46.

[63] See Macrory Consultation *Ibid* paras 3.52–3.71, DEFRA Review (above n 9) paras 6.3.12–6.3.18 and Henley Centre Headlight Vision, *Report on administrative penalties simulations* (May 2006), commissioned by DEFRA as part of the environmental enforcement review process and available at: <http://www.defra.gov.uk/environment/enforcement/pdf/envleg-enforce-henleyreport.pdf> accessed 5 January 2009.

[64] See s II D above.

ground. Although this would add to the costs associated with the MAP process, it would serve to ensure inter alia the consistent application of penalties across a regulated sector and the presence of some form of internal negotiation between the regulator and regulatee prior to a variable MAP being imposed. It would also strengthen transparency and accountability within the system, protect industry from possible abuses of power, and reduce the likelihood of costly appeals and reviews. Again, this is the case in Ontario where the penalty under a variable MAP is determined by a Director of the Ministry of Environment. An important principle emphasised by Macrory is that any revenue generated by MAPs (both fixed and variable) is not accessed (either directly or indirectly) by the regulator that imposed it.[65] The principle would not necessarily prevent money from being ring-fenced for particular purposes. This might, in fact, be desirable, if as suggested in the DEFRA review, MAPs are seen not only as a means of deterring non-compliance but also of securing remediation of damage and making restitution to adversely affected communities.[66] Rather, that any money be managed independently from the regulatory authority that was responsible for collecting it.

In July 2008, the Regulatory Enforcement and Sanctions Act received royal assent.[67] Part III of the Act provides for 'civil sanctions' (administrative as opposed to judicial) such as fixed monetary penalties, and discretionary requirements, including monetary penalties. Regulators can select which, if any of the sanctions, to introduce into the regulatory regimes for which they have responsibility.[68] Under section 39 of the Act, where the regulator is satisfied beyond reasonable doubt that a person has committed a relevant offence, the regulator can impose a fixed monetary penalty (normally capped at £5,000). Details of fixed penalties will be provided for by the Minister in question. They may be set at a single amount, or they may vary according to certain factors such as the size of the business.[69] In terms of process, under section 40, where a regulator proposes to impose a fixed MAP, it must serve a notice of intent which gives the proposed recipient an opportunity to: (a) discharge liability for the fixed MAP by payment

[65] Macrory Consultation (above n 9), para 3.62.

[66] DEFRA Review above n 9 para 6.3.12.

[67] For details of the Act, associated consultations and parliamentary debates see: <http://www.berr.gov.uk/bre/inspection-enforcement/implementing-principles/sanctions-bills/page44047.html> accessed 5 January 2009.

[68] The consultation procedure which must be followed if a regulator or government department wishes to introduce administrative penalties into a particular regulatory regime, was severely criticised by Baroness Young, Chief Executive of the Environment Agency during a reading of the Bill in the House of Lords. She argued that '[T]he hurdles that are now in the Bill—I am referring to Part 3 on access to the new flexible penalties—are so high that they are almost unscaleable. First, regulators have got to be Hampton-compliant. We all want to be Hampton-compliant, so I am sure that is not a hard job. But they then have to get their sponsor body to agree that they should have access to the penalties; they have got to get cross-government departmental approval; the matter has to go to consultation and, if there is any significant change in what has been proposed, it has got to go to reconsultation. To be frank, at the end of all that one might well lose one's will to live.' Hansard Debate HL col 1258 (28 November 2007).

[69] Department for Business, Enterprise and Regulatory Reform, *Regulatory Enforcement and Sanctions Act 2008—Guidance to the Act* (July 2008), para 36.

of a prescribed sum; or (b) make written representations or objections to the regulator.[70] In the latter case, the regulator must decide whether to impose the MAP and if it does, the recipient can appeal to an independent tribunal. Unpaid penalties are enforced via the civil courts and where a fixed MAP is imposed on a person, that person cannot be convicted of the relevant offence for which the penalty was levied.[71]

The Act also provides for the imposition of discretionary requirements which includes a requirement to pay a variable monetary penalty.[72] Guidance would refer to factors to be taken into account in setting variable MAPs. Procedural requirements are very similar to those established for fixed MAPs. One of the key differences is that the person on whom a notice of intent is served can offer an undertaking as to other action to be taken.[73] For example, a business may commit to paying compensation to persons affected by the offence in question and the variable MAP could be reduced to take into account the compensation offered. Parliamentary debates relating to Part III of the Bill produced some interesting responses. Whilst a number of Commons and Lords members expressed support for the provisions in Part III, there was a general concern that the sanctions would be used indiscriminately to punish businesses.[74] Questions were also asked about how and whether regulators could cope with implementing such a complex penalty regime.[75] What was made very clear was that prosecution would remain the right course of action for serious offences. The Minister for Employment Relations and Postal Affairs went on to state that the government did not want 'to criminalise well-meaning businesses unnecessarily, but [did] of course reserve that option for the more serious and consistent offenders.'[76]

The provisions of Part III only came into force on 1 October 2008, so it is very difficult to say how the penalty provisions will be used and to what extent they will deter individuals and companies from polluting. This is of course presuming that

[70] Concerns were expressed that this would make the regulator both judge and jury. See eg Hansard Debate HL col 1278 (28 November 2007) Lord De Mauley: 'The Bill also suggests, as I read it, that if a monetary penalty is imposed on a business, that business can make written submissions only to the very regulator that intends to impose the penalty. That regulator thus becomes, as the noble Lord, Lord Razzall, said, policeman, judge, and jury. There is no chance to question the evidence or enter a plea and have the case heard by an independent person. If the regulator confirms the penalty, while an appeal is possible, it is an appeal by a person already convicted, so it is on the basis of the need to prove innocence, rather than of the accepted British custom of the need for the enforcer to prove guilt. That person or business is also deprived of its right to go to court to secure justice, the decision on whether to pursue this route being entirely in the hands of the enforcer. As the noble Lord, Lord Razzall, argued, this legal anomaly has a deeply prejudicial effect on small businesses. If accused of misconduct, they, unable or unwilling to pay the sizable costs of defending themselves in court, will find themselves being forced to pay the imposed fine with no real option of contesting it. Is the Minister really content to let such a discriminatory appeals system be introduced into the regulatory arena?'

[71] S 40.

[72] S 42.

[73] S 43(5).

[74] See, eg, comments of Graham Stuart MP, House of Commons Second Reading of Bill, 21 May 2008, col 332–3.

[75] See, eg, Baronnes Hale, Hansard Debate HL col 1258 (28 November 2007).

[76] House of Commons, Second Reading of Bill, col 332 (21 May 2008).

DEFRA and the Environment Agency adopt the monetary penalty powers made available to them in the Act. Clear guidance will have to be drafted to ensure that the process is transparent and that variable MAPs in particular are used fairly and are proportionate to the offence committed. It will also be necessary to establish a fast and efficient review/appeal process. Providing that these issues can be dealt with satisfactorily, I believe that the availability of monetary administrative penalties will undoubtedly strengthen the enforcement toolbox and give Agency officers another weapon with which to fight non-compliant individuals and companies.

III Non-monetary Administrative Sanctions

A Administrative Notices and Orders

i Introduction

Where an enforcement officer has reason to believe that a pollution control offence has or is likely to occur, the officer may serve what are referred to as inter alia administrative orders, statutory enforcement notices or directions. The range of notices available to a regulator will vary although common examples include notices that specify the steps which must be taken to come into compliance and the timescale within which changes must be implemented, notices which require a specific activity to be suspended until remedial action is taken and notices that actually prohibit an activity from taking place.[77] In many instances, non-compliance with a notice constitutes a criminal offence.

These devices are generally found towards the base of the regulatory enforcement pyramid, above advice and formal warnings, but below monetary administrative penalties, criminal prosecution and licence suspension/revocation. They are, however, an extremely effective way of dealing with the recalcitrant individual or company and represent a 'light touch approach to enforcement, which does not unnecessarily punish businesses that are willing to take swift action to come into compliance'.[78] From a deterrence perspective, administrative notices and orders send a strong deterrence message. First, complying with a notice will impose costs on the recipient; costs which would not otherwise have been borne. Second, compared with the cost of prosecution, the procedural costs associated with the serving of such notices are relatively low and so, like MAPs, they are more likely to be utilised than a criminal prosecution. And third, their deterrent impact is strengthened by the fact that non-compliance with a notice often, in itself, constitutes a criminal offence and can result in prosecution.

[77] Specific examples are given in ss ii–iv below.
[78] Macrory Consultation (above n 9), para 4.12.

They may also be as, if not more effective than criminal prosecution for a number of other reasons. Like MAPs, both fixed and variable, administrative notices can be used without intervention by a court or tribunal. They therefore enable an enforcement officer to respond quickly and positively to both actual and imminent environmental harm caused by polluting activities. By focussing on future compliance, administrative notices also reflect a risk-based approach to regulation, aim to change behaviour and may move the recipient into compliance. Furthermore, decisions as to the serving and content of these notices are made by regulators with expertise in addressing environmental problems, as opposed to judges with no specialist knowledge of environmental matters. This is particularly relevant in those jurisdictions without a specialist environmental court or tribunal. Notices and orders are also arguably more effective than financial penalties, both administrative and judicial. Monetary penalties may not have the desired effect: a small firm may not be able to pay a fine, and for the larger firm, monetary penalties may be seen as a 'cost' of doing business.

All in all, administrative notices are a vital component of the environmental regulators' enforcement toolbox and are important in both deterring non-compliance and promoting future compliance. This is reflected in the extensive provision that is made for such notices and the fact that in many jurisdictions, notices are a more frequent response to non-compliance than criminal prosecution.

ii Administrative Notices in Australian Pollution Control Regulation

Administrative notices are a key tool in the regulator's enforcement ammunition and will be issued where substantive work or action is required to comply with regulatory provisions. Recent legislative amendments to pollution control legislation in Victoria and NSW strengthen administrative notice provisions and are testimony to the important role they play in deterring non-compliance. Although the conditions under which a notice can be served and the precise terms of that notice are governed by legislative provisions, a number of general observations can be made.

First, notices may be positive (requiring the recipient to do something), negative (restraining the recipient from acting in a certain way), or both. For example, in Victoria, EPA officers can serve a clean-up notice on inter alia the person who caused or permitted the pollution.[79] Alternatively, on being satisfied that a process or activity has caused or is likely to cause pollution, an enforcement officer can serve a pollution abatement notice requiring the recipient to inter alia cease carrying on the activity or modify or control the process in some way.[80] Legislation in NSW makes provision for a number of administrative notices. For example, a prohibition notice can be served where the regulator is of the opinion that the emission or discharge of pollutants from premises is causing or is likely to cause,

[79] EPA1970 s 62A (Victoria).
[80] EPA 1970 s 31A (Victoria).

inter alia, harm to human health or the environment[81] and a prevention notice can be utilised where an activity is being carried out in an environmentally unsatisfactory manner.[82] In some states, such as Victoria and Queensland, where there is an imminent threat to the environment or human health, an authorised officer may issue directions to, for example, remove, disperse, destroy or treat any pollutant, waste or substance.[83] As these directions are reserved for emergencies, they will normally be given verbally and then confirmed in writing shortly afterwards. Like administrative notices, directions can impose both positive and negative requirements on the recipient.

A second key characteristic of administrative notices in Australian pollution control regulation is that they may be accompanied by other enforcement measures such as an on-the-spot fine or, for more serious offences, prosecution. Third, the notice will give recipients a particular time-frame in which to comply with its requirements, taking into account the nature of the work involved and the likelihood and seriousness of any environmental harm that may be caused if the work is not carried out.[84] Fourth, many regulators are given the authority to recoup administrative costs involved in the drafting and serving of a notice. For example, section 60C of the EPA 1970 (Victoria) requires that the Authority impose a service fee of 40 fee units (currently $454 in total) on recipients of a pollution abatement notice.[85] In addition, the recipient of a notice may be liable for any monitoring or compliance costs incurred by the regulator.[86] Finally, administrative notices and directions are under-pinned by criminal liability: in other words, non-compliance will commonly constitute a criminal offence.

The power to issue such notices and directions is not unfettered. Any notice, order etc should reasonably relate to the activity in respect of which it is issued. Furthermore, it is imperative that the provisions of the notice are clear enough so as to enable the recipient to understand what action must be taken. As non-compliance attracts criminal liability, this is particularly important. For this reason, legislation which confers on regulators the power to serve an administrative notice will often require that they provide reasons for doing so and that they clearly stipulate the measures needed to ensure compliance.[87] For example, in *Re Robert Lawrence; ex parte Goldbar Holdings Pty Ltd* the Supreme Court of Western Australia quashed some parts of a pollution abatement notice on the basis that they were ultra vires for uncertainty and for invalid delegation of powers.[88] The notice purported to tackle dust and spray emanating from the operation of a

[81] See, eg, PEOA 1997 (NSW) ss 101–3.

[82] PEOA 1997 (NSW) ss 95–100.

[83] See, eg,EPA 1970 (Victoria) s 62B .

[84] In some cases, legislation will specify a minimum time for compliance. For example, in Victoria the minimum time for compliance with a pollution abatement notice is 30 days.

[85] As applicable from July 2008.

[86] See, eg, PEOA 1997 (NSW) ss 104–7. Only a Minister can issue a prohibition notice, on the advice of the EPA (as was) or the Sydney Catchment Authority.

[87] For example, EPA 1970 (Victoria) s 62A.

[88] (1994) 84 LGERA 113.

limestone cutting machine operated by the defendant company, and dust caused by vehicular traffic from an unsealed access driveway. Although the relevant section of the Environment Protection Act (Western Australia) required the Chief Executive Officer to 'specify' the reasons for serving the notice and the measures which had to be taken, it was somewhat ambiguous in its requirements, stating, for example, that measures had to be taken to the 'satisfaction of the inspector' or that 'other measures' could be taken providing they were approved by the Director.[89] The affected provisions were therefore quashed. The approach of the court in this case is similar to that of the Victorian Supreme Court in *EPA v Simsmetal Ltd*, in which a notice was deemed invalid on the basis that it did not specify clean-up measures and therefore the recipient could not have known how to comply with the notice.[90] According to Gobbo J:

> A vague direction to consider a variety of general matters is almost impossible to enforce for it will be very difficult to prove that the matters were not considered unless the recipient admits that it did not do so.[91]

Cases such as *Goldbar Holdings* and *Simsmetal* have undoubtedly made environmental regulators more cautious in their approach to drafting administrative notices and orders. The need to provide sufficiently clear detail on the measures to be taken under the notice will have raised the cost to regulators of utilising notice powers, arguably making them less cost-effective.

A perusal through the Annual Reports and Accounts of some Australian environmental regulators will illustrate that, in comparison with the use of criminal prosecution, administrative notices are commonly used by enforcement officers faced with actual or potential breaches of the law. For example, in contrast with the 13 major prosecutions brought for pollution offences, the Victorian EPA, in 2006–07 issued 197 pollution abatement notices and 44 clean-up notices. Similarly in 2005–06, compared with 12 'major' prosecutions, the EPA issued 239 pollution abatement notices and 52 clean-up notices.[92] No statistics are available for the issuing of administrative notices in NSW.

iii Administrative Notices in Canadian Pollution Control Regulation

A range of administrative notices are available under both Ontario's pollution control laws and CEPA. Many of these notices mirror those available in Victoria and NSW in terms of effect, and display similar characteristics. In all cases, they can be used in conjunction with other enforcement tools such as written warnings and prosecution and as such form an important part of the enforcement toolbox.

[89] See further, A Harman, 'Pollution Abatement Notices: The Requirement for Certainty' (1995) 12 *Environmental and Planning Law Journal* 147.
[90] (1990) 70 LGRA 312.
[91] (1990) 70 LGRA 312, 322.
[92] See EPA (Victoria), *Annual Report 2006–07*, App.

In Ontario 'control documents' (such as control, stop and remedial orders)[93] can be issued in the event of a breach. Although such documents do not require the 'offender' to pay a monetary fine, compliance with the document will require time and money to be invested in ensuring future compliance. The authority to issue such documents is vested by statute in a Director of the Ministry of Environment. However, to aid effective and efficient inspection and enforcement, the powers have been delegated to provincial officers, with some limitations.[94] Officers also have the power to serve a Provincial Officer's Order on a recalcitrant individual or company.[95] A Provincial Officer's Order can be served when the officer reasonably believes that a provision of the EPA or regulations, a provision of an order or a term or condition of a certificate of approval has been contravened. The orders can prescribe a wide range of conduct, from preventing the continuation or repetition of a contravention to repairing any injury or damage to the environment. In the event that environmental damage is likely to occur, the order can require the recipient to prevent such damage.

There are no general statistics on the use of control documents or indeed Provincial Officer's Orders. However, there is some evidence which may suggest that Provincial Officer's Orders are seen as an extremely important abatement tool. From December 2000 to July 2001, Ontario's Environmental SWAT team (now the Sector Compliance Branch), which focuses on sectors with high non-compliance rates and/or significant public health and environmental impacts, inspected 61 electro/metal plating businesses in Ontario.[96] Approximately 85 per cent of the businesses inspected required some enforcement action to increase compliance. Seventy Provincial Officer's Orders were issued in total, compared with nine tickets under the Provincial Offences Act. Seven cases were referred to the Ministry's Investigations and Enforcement Branch which decides whether or not to lay criminal charges. Therefore, in most of these cases, non-compliance was dealt with by way of a Provincial Officer's Order and by November 2002, 95 per cent of companies had complied with the orders issued.

At federal level, CEPA also prescribes a number of administrative notices, non-compliance with which is a criminal offence.[97] Any offence under CEPA can be dealt with by an Environmental Protection Compliance Order (EPCO) an administrative tool which is equivalent to the administrative orders and notices available to Ministry of Environment officials in Ontario. An EPCO can only be applied where enforcement officers have reasonable grounds to believe that a contravention of the Act has or will occur in relation to a range of prescribed activities which

[93] Control orders: EPA 1990 (Ontario) ss 7, 124–8. Stop orders: EPA 1990 (Ontario) ss 8, 128. Remedial Orders: EPA 1990 (Ontario) s 17.

[94] See Ministry of Ontario Compliance Guidelines Schedule 2. Where the power to issue a notice is delegated to provincial officers, the document is termed a 'field order'.

[95] EPA 1990 (Ontario) s 157.

[96] See 'Environmental Compliance in the Electro/metal Plating Sector: An Inspection Sweep by the Environmental SWAT Team' available at <http://www.ene.gov.on.ca/programs/4292.htm> accessed 5 January 2009.

[97] See generally CEPA Pt 10.

are, on the whole, related to substances government by the Act.[98] The order can inter alia require an immediate stop to the illegal activity or require that corrective action be taken.[99] Unlike the administrative notice powers examined so far, under CEPA, regulatees have the right to respond orally or in writing to a draft EPCO. Any such representations will be taken into account by the enforcement officer issuing the EPCO and a draft order can be modified, issued unchanged or not issued at all. The EPCO process therefore clearly involves an element of negotiated compliance. Whilst this may advantageous in the sense that the process will result in an order which is tailored to the offence and the offender, it also adds to the time and cost associated with the notice system and may reduce the likelihood of an EPCO being served. This may weaken the deterrent impact of EPCOs. Saying that, they are clearly viewed as valuable enforcement tools, with 100 EPCOs issued in 2004–05, compared with 13 prosecutions.[100] Statistics on enforcement of CEPA in previous years do not specify the number of EPCOs issued. In the event of a release of substances in contravention of regulations under CEPA 1999, enforcement officers can also issue 'directions' which require the recipient to take all reasonable emergency measures to inter alia prevent the release if it has not already occurred, or remedy any dangerous condition. Ministerial orders such as prohibition and recall orders can also be used to respond to certain alleged violations. As is apparent in their title, these orders can only be issued by the Minister of Environment, not by enforcement officers.

iv Statutory Notices in England and Wales

In England and Wales, legislation makes provision for the serving of a range of administrative notices, non-compliance with which constitutes a criminal offence. For example, where a site breaches or is likely to breach the conditions of its environmental licence, Environment Agency officers can serve an enforcement notice on the offender, specifying any steps which must be taken to remedy the breach.[101] And in the case of a potential or actual water pollution incident, the Agency can serve a works notice to prevent or remedy the pollution and can recover the costs of the investigation.[102] Non-compliance with an administrative notice will normally be a criminal offence.[103] It is very difficult to obtain statistics relating to the use of administrative notices. However, data would suggest that the Agency uses notices much less frequently than it does criminal prosecution. For example, in

[98] See CEPA ss 234–71.

[99] See CEPA ss 234–242.

[100] CEPA 1999, *National Enforcement Statistics 2004–05* available at <http://www.ec.gc.ca/ele-ale/default.asp?lang=en&n=0306B67B>accessed 5 January 2009.

[101] See eg s 13 EPA 1990 and regulation 24 PPC Regulations 2000. In addition, where a process is being carried out in such a way that there is an imminent risk of serious pollution, the Agency can serve a prohibition notice, requiring the recipient to carry out certain steps to remove that risk. See, eg, EPA 1990 s 14, and PPC Regulations 2000 reg 21.

[102] Anti-Pollution Works Regulations 1999 SI 1999/1006 reg 2 .

[103] See eg EPA 1990 s 23 and PPC Regulations 2000 reg 32.

2004, the Agency served 563 notices compared with 740 prosecutions. This was considerably higher than the 396 notices served in 2003.[104]

All in all, the current system of enforcement notices in environmental law has received much support. According to evidence submitted to the Macrory Review, a government-commissioned review that examined regulatory sanctions, 52 per cent of those who responded to a discussion paper of December 2005 on regulatory justice, felt that the current system achieved its goal, with only 11 per cent of respondents saying that it did not.[105] There is, however, room for improvement. Macrory found, for example that the value of enforcement notices as a sanctioning tool is weakened by the fact that some regulators appear not to follow up enforcement notices once they are served. Statistics as to the number of prosecutions launched as a result of non-compliance with an administrative notice are not available. But what is clear is that failure to monitor compliance with these notices will mean that they will not be taken seriously enough by the regulated community. Macrory suggested that one way of strengthening the sanctioning impact of these notices is to empower regulators to serve a monetary administrative penalty on an offender who has failed to comply with a notice.[106] Another problem identified by DEFRA was that there is no general power to issue a 'stop notice' where, for example, there is a serious risk of environmental harm. Such a power would, DEFRA argued, help the Environment Agency 'in particular to shift effort to deal more efficiently and effectively with those least likely to stop when asked.[107]

The government responded to such concerns in the Regulatory Enforcement and Sanctions Act 2008. The Act introduces enabling powers which, if adopted by DEFRA, would give Environment Agency officers: (a) the power to serve a stop notice on a person; and (b) the power to use 'discretionary requirements' against a person who has committed a relevant offence.[108] Stop notices under sections 46–49 of the Act could be used where the regulator reasonably believes that the activity being carried on by that person is causing or presents a significant risk of causing serious harm to inter alia the environment. Non-compliance with the notice would constitute a criminal offence.[109] Discretionary requirements include variable monetary penalties (discussed in Section A above), requirements to take such steps so as to secure that the offence does not continue or recur (such as repairing a piece of broken equipment, changing a process or providing training), and requirements to restore the position to what it would have been had the offence not been committed (such as cleaning up pollution).[110] Where a regulator is given all three powers, they can be used in combination and tailored to reflect

[104] ENDS Report, 'Agency to target 'big, bad and nasty' crime' (May 2005) 364.

[105] Macrory Consultation (above n 9), 40. Respondents include industry representatives, regulatory scholars and public interest groups.

[106] Macrory Consultation (above n 9) para 4.21 and 4.22. This may well be introduced under the fiexed/variable MAP provisions of the 2008 Act.

[107] DEFRA Review (above n 9), para 6.3.23.

[108] As defined in s 38 of the Act.

[109] For guidance on stop notices under the Act see above n 69, paras 51–7.

[110] See s 42(3).

the heterogeneity of the offence and the offender. According to the Department for Business Enterprise and Regulatory Reform:

> The discretionary requirements will give regulators a flexible set of powers that enables them to achieve a constructive enforcement outcome that remedies the consequences of an offence and promotes good working relationships between the regulator and business; an outcome that is often not served by prosecution.[111]

The procedure for imposing compliance and restoration-based discretionary requirements is the same as that for serving variable monetary administrative penalties. In other words, a notice of intent must be served on the person committing the relevant offence, and the proposed recipient can then make representations. This consultation procedure is similar to that which must be followed if Environmental Protection Compliance Orders are issued under CEPA (Canada). However, as non-monetary discretionary requirements are less punitive in nature than variable monetary penalties, a business who fails to comply with such requirements can be criminally prosecuted at a later date for the original offence.[112] Alternatively, the regulator may be able to issue a monetary penalty for non-compliance.[113] This, of itself, gives effect to Macrory's recommendation and may go some way to improving compliance with administrative notices.

B Licence Suspension and Revocation

i Introduction

Where the alleged offender operates under an environmental licence, regulators may also have the option of revoking or suspending all or parts of that licence. The revocation or suspension of a licence can close down a factory, or severely limit the output of a manufacturing process. By withdrawing some or all of the operating rights under a licence, suspension or revocation can deal a more severe economic blow than prosecution and is correspondingly found at the pinnacle of the enforcement pyramid. It is therefore arguably a more punitive sanction than prosecution and sends a very strong deterrence signal. However, the deterrent impact of suspension and revocation is, unsurprisingly, weakened by the fact that regulators are extremely cautious in using these powers. In addition to deterrence, suspension and revocation also clearly serve a preventative function and are therefore defensible on the grounds of public protection. They are particularly valuable where the 'social harm arising from the contravention is very large and/or there are problems in deterring the conduct by *ex post* sanctions.'[114]

[111] Above n 69, para 44.
[112] See s 44(3).
[113] S 45.
[114] S Shavell, 'The Optimal Structure of Law Enforcement' (1993) 36 *Journal of Law and Economics* 255, 261–2.

ii Licence Suspension and Revocation in Australian Pollution Control Regulation

Powers of revocation and suspension are common in Australian pollution control legislation.[115] However, compared with other administrative enforcement tools, the power to suspend or revoke a licence is used sparingly and will be reserved for cases where, for example, the licensee has a history of breaching licence conditions or the unlawful conduct continues to occur even after criminal prosecution. The EPA Victoria Enforcement Policy goes so far as to say that 'revocation is a serious step that the EPA does not contemplate lightly.'[116] An officer in the EPA Victoria stated that they are used predominantly as threats, whereby the 'offender' will receive a notice of possible revocation or suspension.[117] Similar sentiments were expressed within the NSW EPA where reference was made to the fact that alternative, less draconian measures such as serving an environmental notice or applying for an injunction could be used with similar results.[118] To provide a sense of how these powers are used, in NSW licence revocation and suspension is a two stage process: full or partial revocation and suspension is pre-empted by the issuing of a notice of intention to proceed with such action. The NSW EPA in 2002/03 issued only five notices of intention to suspend which led to three licence suspensions. With regard to revocation, the EPA served 14 notices of intention to revoke, leading to eight licence revocations.[119]

iii Licence Suspension and Revocation in Canadian Pollution Control Regulation

The Ontario EPA also empowers the Director to suspend or revoke a certificate of approval.[120] Described as a power of 'last resort', there are very few examples of it being exercised.[121] In most cases, such an order is issued in conjunction with various control documents of the type mentioned in section B above. For example 310 Waste Ltd's Certificate of Approval to operate a waste transfer facility was suspended in October 2004 and an order requiring the owners and operators of the site to remove and dispose all of excess waste was also served. In addition, the owners of the site faced criminal charges.[122] Under section 140 of the Act, a person to

[115] See, eg, PEOA 1997 (NSW) ss 79–82 and EPA 1970 (Victoria) s 20. In both cases, the licence holder can appeal the suspension or revocation, to the Land and Environment Court (NSW) and the Victorian Civil and Administrative Tribunal (Victoria). EPBCA (federal) ss 144, 145 also give Ministers the power to suspend or revoke an environmental approval issued under the Act where, eg, the Minister believes on reasonable grounds that a significant impact on matter protected by the provision has occurred because of a contravention of a condition attached to the approval.

[116] EPA (Victoria), *Enforcement Policy* (July 2006) 13.

[117] During interviews carried out in March 2004.

[118] During interviews carried out in March 2004.

[119] Information provided during interview with EPA officers in March 2004.

[120] See eg s 39(2) EPA 1990 (Ontario).

[121] During an interview with a Ministry of Environment inspector.

[122] See Ministry of Environment News Release October 25 2004 and November 24 2004.

whom an order is directed may appeal to the Environmental Review Tribunal and from there to the Divisional Court on a question of law. Licence suspension and revocation are not available when dealing with alleged violations of CEPA.

iv Licence Suspension and Revocation in Pollution Control Regulation in England and Wales

In England and Wales, Environment Agency officers can suspend or revoke environmental licences where, for example the continuation of the activities authorised by the licence would cause pollution of the environment and this cannot be avoided by modifying the conditions.[123] The potentially severe economic impact of licence suspension or revocation is reflected in the fact that compared with the serving of administrative notices, these powers are rarely used. For example, in the first five years of the Agency (1996–2001), only one waste management licence was revoked. However, since that time, numbers seem to be rising. According to information provided by the Environment Agency, a total of 35 revocation notices were served from April 1999 to May 2008, 33 of which related to waste management licences. The number of suspension notices served is, unsurprisingly much higher. During the same time period, a total of 370 suspension notices were served. 361 related to waste, seven to IPPC and two to water quality.

C Environmental Audits

One particularly interesting example of an administrative enforcement option currently available to some environmental regulators in Australia is the environmental audit order. As was discussed in Chapter 7, the primary purpose of such orders is arguably not to punish or deter the offender, but to improve environmental performance and possibly prevent the recurrence of an offence.[124] An audit, by providing the regulator with a comprehensive and objective analysis of whether environmental requirements are being met, improves monitoring and allows enforcement action to be targeted at the most risky aspects of the regulated activity. And in addition to preventing a recurrence of the original offence, an audit can be useful where it is likely that a violation will be repeated at other operations owned by the company who committed the initial offence. By requiring the company to audit all, or some of its sites, it is hoped that any potential violations will be identified and corrected before harm is done. Mandatory environmental audits are, in fact, especially useful in regulating 'environmental laggards' that have a very poor environmental record.[125] There are, however, two clear punitive,

[123] EPA 1990 s 38. Appeals against licence suspension/revocation are made to the Secretary of State (s 43).

[124] See specifically s VI E.

[125] N Gunningham and D Sinclair, 'New Generation Environmental Policy: Environmental Management Systems and Regulatory Reform' (1998) *Melbourne University Law Review* 592, 595–8.

deterrent aspects to the sanction: the company must fund the cost of the auditing procedure and the results must be disclosed to the regulator.

As discussed previously, some jurisdictions give the courts the power to impose an environmental audit order. However, the power is arguably better vested in regulators who can respond quickly and effectively to future risks of non-compliance. Pollution control regulation in Australia often makes provisions for administrative (as opposed to judicial) environmental audit orders. Circumstances in which an audit may be required include where the regulatory authority suspects that a person has breached the relevant Act, regulations, licence conditions or administrative orders.[126] The position in NSW is slightly different, with mandatory audits being permitted only where provision is made for such action in a licence condition.[127] At federal level, the EPBCA gives the Minister the power to direct that an environmental audit be completed where the environmental impacts of the authorised activity are greater than was anticipated when the activity was first assessed, or that the holder of the authority is likely to breach a condition of that authority.[128]

Mandatory audits cannot be required under either the EPA (Ontario) or CEPA (Canada federal). Similarly, pollution control regulation in England and Wales does not make provision for mandatory environmental audits and it would seem that the Environment Agency does not use licence conditions to require such action. Although not explicitly referred to in either the Macrory or DEFRA review process the enforcement hand of the Environment Agency would undoubtedly be strengthened if officers could compel a company to carry out an environmental audit.

D Enforceable Undertakings

i Introduction

Enforceable undertakings are an additional and rather innovative enforcement tool that has recently been introduced in several jurisdictions including Australia and England and Wales. They are similar to administrative notices in that they are alternatives to paying a fine.[129] The key difference is that they constitute a negotiated agreement between the regulator and offender, in which the offender commits to take specified actions relating to non-compliance. The agreement will be tailored to the individual circumstances of the offence and the offender, and can typically require the offender to remediate environmental harm, compensate affected parties, instigate an environmental audit, correct deficient operating pro-

[126] See, EPA (Victoria) Pt IXD and Environmental Management and Pollution Control Act 1994 (Tasmania) s 30.

[127] PEOA 1997 (NSW) s 67.

[128] See EPBCA (federal) s 458.

[129] For an interesting summary of issues relevant to enforceable undertakings see ALRC, *Civil and Administrative Penalties in Australian Federal Regulation* (Report 95, 2003) ch 16.

cedures and carry out a community project. By combining aspects of a number of other administrative enforcement options, enforcement undertakings therefore represent a flexible and effective approach to enforcement and compliance

Enforceable undertakings have a number of additional benefits. First, they will invariably be used where there has been a contravention or alleged contravention of statutory provisions. However, instead of (or as well as) proceeding with cumbersome and costly criminal prosecution, the regulator and offender can agree to enter into an undertaking, non-compliance with which constitutes an offence. There are, of course, significant administrative costs involved in negotiating a settlement agreement; they are not a 'quick fix', with drafting taking considerable time and effort for both parties. Nevertheless, they do provide:

> [A] quicker and more cost-effective mechanism for resolution than court proceedings [and amount to] an efficient and effective use of resources.[130]

Second, by taking into account industry resources, they avoid the 'deterrence' trap strongly associated with criminal fines. And third, unlike other administrative tools such as administrative notices, they are arguably more appropriate for serious offences and can impose very significant financial and indeed reputational costs, by virtue of them being made publicly available. Parker also identifies, as important, the restorative justice element of enforceable undertakings. If used appropriately, she argues that they can constitute

> [A] form of alternative dispute resolution in which a potential conflict about legal liability is transformed into a negotiated resolution whereby all the relevant parties agree to make sure justice is done.[131]

Contrary to claims that they are a soft option which allow businesses to avoid criminal prosecution and the subsequent court-imposed penalties, Parker submits that:

> [Enforceable undertakings] can deliver superior identification and accountability of businesses and their managers, and also other individuals, institutions and systems responsible for breaches of the law then enforcement action through litigation often does.[132]

It is, of course, important to recognise that enforceable undertakings must not be used by regulatory authorities to 'bully' businesses into complying with laws. The system must be subject to procedural safeguards which protect important process values such as transparency, consistency and accountability. Discretion must be checked to ensure that they resemble lenient, non-punitive versions of what a court might have ordered. Provided that there is clear guidance in place, enforceable undertakings should therefore be seen as an important addition to the regulatory enforcement toolbox. They should not, however, be used to the exclusion of litigation.

[130] ALRC Report (above n 12) para 16.57.

[131] C Parker, 'Restorative Justice in Business Regulation? The Australian Competition and Consumer Commission's Use of Enforceable Undertakings' (2004) 67 *MLR* 209, 213.

[132] *Ibid* 212.

Although new in the environmental context, enforceable undertakings have, for some years, been part of the enforcement model adopted by a number of Australian regulatory bodies including the Australian Competition and Consumer Commission (ACCC), the Australian Securities and Investments Commission (ASIC) and the Victorian WorkCover Authority.[133] An investigation into the use of enforceable undertakings by the ACCC provides a valuable insight into how they can work in practice. The ACCC will only accept an undertaking if there is evidence of non-compliance which would otherwise justify litigation. Factors taken into account in deciding whether to accept an undertaking rather than take court action include whether a penalty is warranted and whether the previous history of the offender suggests that an administrative settlement will be enough to deter future non-compliance.[134] In terms of their use, contrary to published reports, they were issued extensively from mid-1997 to mid-2002. Evidence from annual reports revealed that enforceable undertakings were accepted in 37 per cent of matters in which the ACCC took enforcement action, and in 22 per cent of cases, an undertaking was the only enforcement action reported.[135] So although there was a predominant reliance on litigation, the power to negotiate an enforceable undertaking was used frequently. They may not, of course, be appropriate in all serious cases of non-compliance. An internal review of enforceable undertakings in 1998 revealed that in some cases a court order may also be legitimately sought, particularly where general (as opposed to specific) deterrence is an aim of enforcement action.[136]

So, all in all, enforceable undertakings are extremely useful sanctions that enable regulators to tailor their enforcement response to the individual circumstances of the offence and the offender. This is especially evident in studies relating to their use by the ACCC. Compliance with a settlement agreement will undoubtedly impose financial costs on an offender. If we combine this with the fact that they are more cost-effective than criminal prosecution, then they can send a strong deterrence signal to the regulated community. However, deterrence is arguably not the most important rationale of enforceable undertakings. By including measures which provide for remediation, compensation and rehabilitation, all of which are valuable goals in tackling non-compliance with pollution control regulation, EUs can be more satisfying for both the regulator, the offender and the victims of the offence. The restorative justice element of EUs should also not be underestimated. If used as intended, the outcomes of a restorative justice approach can be significantly better than those achieved through litigation.

[133] For detailed analysis of the enforceable undertakings in Australian law see Parker (*Ibid*), K Yeung, *The Public Enforcement of Australian Competition Law* (Canberra, ACCC, 2001) ch 5 and K Yeung, *Securing Compliance—A Principled Approach* (Oxford, Hart Publishing, 2005) chs 6 and 7.

[134] ACCC, *Submission to the Trade Practices Act Review* #56 (2002) para 10.6.2 cited in Parker (*Ibid*) 216.

[135] Parker (*Ibid*) 216–17.

[136] D Watt, 'Evaluation of the Use of s 87B Undertakings' (1998) 13 *Australian Competition and Consumer Commission Journal* 7, 8, cited in Parker *Ibid* 217.

ii Enforceable Undertakings in Australian Pollution Control Regulation

In NSW and Victoria, enforceable undertakings were introduced in 2005 and 2006 respectively.[137] Sections 67D–67G of the EPA 1970 (Victoria) provide an outline of the system, and the EPA has produced draft guidance on how the enforceable undertakings would work in practice.[138] According to the Act, they can be used where a person has 'contravened or allegedly contravened a provision of any Act or regulation in respect of which the Authority make take proceedings for an offence.'[139] The EPA must also be satisfied that an enforceable undertaking is an appropriate enforcement mechanism, taking into account criteria to be specified in guidelines.[140] These factors include the seriousness of the offence, the culpability of the offender, the cost of enforcement and whether enforcement proceedings may be counterproductive, and the precedent that may be set by any failure to take enforcement action.[141] If an enforceable undertaking is accepted, proceedings cannot be brought for the original offence and if a person complies with the undertaking, no proceedings may be brought for the offence in respect of which the undertaking was given.[142] In the event of non-compliance, the EPA can apply to the magistrates' court for its enforcement.[143]

In NSW, the possibility of introducing court-enforceable undertakings was first discussed in the PEOA review which commenced in 2002. An Issues Paper released in June 2003, in describing court enforceable undertakings as being effective in 'providing an enhanced enforcement capability for other regulators such as the ACCC and the ASIC', invited comments on whether or not the EPA should have the power to accept such undertakings.[144] The review process revealed no major opposition to the introduction of enforceable undertakings and the PEO (Amendment) Act 2005 duly gave the (now) DECC the power to enter into court-enforceable undertakings. The legislative provisions are similar to those under the EPA (Victoria) although where there is non-compliance with an enforceable undertaking, a case can be brought in the Land and Environment Court.[145] Somewhat surprisingly, the DECC does not seem to have issued guidance relating to the use of enforceable undertakings. Despite this, two enforceable undertakings have recently been agreed, the most recent in early 2008, when Delta Electricity

[137] NSW: PEOA 1997 s 253A, inserted by Protection of Environment Operations (Amendment) Act 2005 No 96 Sch 1. Victoria: EPA 1970 ss 67D–67G, inserted by Environmental Protection (Amendment) Act 2006. For a discussion of enforceable environmental undertakings in Australia see S Streets, P Kruzas and M Payton, 'Environmental Undertakings: Ownership and Obligations' (2002) 19 *Environmental and Planning Law Journal* 180.

[138] EPA (Victoria), *Enforceable Undertakings: Draft Guidelines* (August 2008).

[139] S 67D(1)(a).

[140] S 67D(1)(b).

[141] Above n 138, App 1.

[142] S 67D.

[143] S 67E(2).

[144] NSW EPA, *Issues Paper: A Review of the Protection of the Environment Operations Act 1997* (June 2003) 9.

[145] S 253A PEOA 1997 (NSW).

offered one following the DECCs investigation into a potential pollution incident in February 2007.[146] According to a DECC press release, the company committed inter alia to undertake a voluntary audit, to commission an internal Committee of Inquiry, to install new equipment to undertake staff training, and to fund additional environmental measures costing $45,000. This agreement clearly demonstrates the wide variety of measures that can be agreed to using this power.

Sections 486DA and 486DB of the Environment Protection and Biodiversity Conservation Act (EPBCA) 1999 empower the Director to accept enforceable undertakings where there has been a breach of the civil penalty provisions of Part 3 of the Act.[147] Unlike enforceable undertakings available in Victoria and NSW, an undertaking given under the EPBCA can only relate to the payment of money to either the Commonwealth or to some other person, for the purpose of activities directed towards the protection and conversation of the matter protected by the civil penalty provisions. If the undertaking is not complied with, then the Minister can apply to the Federal Court for an order which directs the person to comply with the undertaking or for 'any other order that the Court considers appropriate.'[148] Information relating to the number of enforceable undertakings agreed to under the Act is not available.

iii Enforceable Undertakings in the UK

Until recently, enforceable undertakings were not available in England and Wales.[149] However, there has been strong support for their introduction for several years. Macrory, in his final report, provided some guidance on how enforceable undertakings could work in the UK. He emphasised the need for transparency and accountability, the importance of guidance on the use of enforcement undertakings and the need for proportionality in what can be demanded. He also advocated the introduction of 'undertakings' plus which would incorporate a financial penalty into an enforceable undertaking.[150] This would be appropriate where, for example, the offender clearly profited from the offence. All conditions of the enforcement undertaking, including the financial penalty, would have to be agreed by both the company and the regulator and the level of financial penalty would be assessed under principles similar to those used in assessing variable monetary administrative penalties.[151]

In recognition of the role the enforceable undertakings can play in regulatory enforcement, section 50 of the the Regulatory Enforcement and Sanctions Act

[146] DECC Media Release, 'Delta Electricity to carry out $45,000 of environmental works under new enforceable undertaking powers' (28 February 2008).

[147] Under Pt 3 of the Act, environmental approvals are required for certain matters.

[148] EPBCA s 486DB(2).

[149] Enforcement orders available to the Office of Fair Trading are similar to enforceable undertakings, but with some key differences. For a brief summary see Macrory Final Report (above n 9), 63–5.

[150] See Macrory Consultation paras 4.41–4.43 and Macrory Final Report para 4.30 (above n 9).

[151] See s II E.

2008 introduces what are called 'enforcement undertakings'. The legislative provisions strongly mirror those available in the EPA 1970 (Victoria) and the PEOA 1997 (NSW). An undertaking can be accepted where the regulator has reasonable grounds to suspect that a relevant offence has been committed. The scope and remit of measures which can be specified in an enforcement undertaking are, however, limited to:

(a) action to secure that the offence does not continue or recur,
(b) action to secure that the position is, so far as possible, restored to what it would have been if the offence had not been committed,
(c) action (including the payment of a sum of money) to benefit any person affected by the offence, or action of a prescribed description.[152]

Where an undertaking is accepted, that person cannot be convicted of the offence to which the undertaking relates, nor can a fixed or variable monetary penalty be imposed on that person. The Act does therefore not make provision for what Macrory calls 'undertakings plus'. Further legislation will be needed to deal with inter alia the procedure for entering into an undertaking, the variation and publication of undertakings and monitoring compliance with undertakings.

iv A Form of Enforceable Undertaking in Canadian Legislation— Environmental Protection Alternative Measures

Federal Canadian environmental legislation, namely CEPA 1999, provides for an interesting and rather innovative alternative to more traditional administrative enforcement mechanisms.[153] Environmental Protection Alternative Measures (otherwise known as EPAMs) allow for a negotiated return to compliance without a court trial. EPAMs are similar to enforceable undertakings discussed above, although the administrative process leading up to the agreement of an EPAM is slightly different. In summary, where a person has been charged with a CEPA offence and that person is willing to take steps to return to compliance without undergoing a trial, the Attorney-General of Canada, in consultation with other interested parties, may negotiate an EPAM with the alleged offender. EPAMs can be used for most violations under CEPA, with the exception of certain offences such as reckless or intentional behaviour that causes an environmental disaster, conduct leading to injury or death of a person and wanton or reckless disregard for human life.[154] Unlike EUs in Australia and England and Wales, although EPAMs are not imposed subsequent to criminal conviction, they are closely connected to the prosecution process. Before an agreement can be reached, a charge for the alleged offence must be laid against the alleged offender and the Attorney-General must be of the opinion that there is sufficient evidence to proceed with the

152 S 50(3).
153 See CEPA 1999 ss 295–309.
154 CEPA 1999 s 296 (1)(b).

prosecution of the offence.[155] Although the accused does not have to plead guilty, he must accept responsibility for the offence.[156] At any time during the process, the accused can elect to have the charge dealt with by the court.[157] By requiring charges to be laid, it is clear that the EPAM process does not necessarily avoid many of the costs associated with criminal prosecution. As a result, it's deterrent impact may be limited. However, as the outcome of enforcement action is determined not by the courts, but by the offender and a regulatory authority with expertise in that particular area of regulation, it will be more acceptable to the offender, and will lead to an agreement which is better designed to meet its goals.[158]

An EPAM may not always be a suitable enforcement response. The Attorney-General must be satisfied that, in the circumstances, an EPAM would be appropriate having regard to the nature of the offence, the circumstances surrounding its commission and a number of other factors. Most importantly, the conclusion of an EPAM must be consistent with the protection of the environment and of human life and health and other interests of society.[159] The Attorney-General should also have regard to the person's history of compliance, whether the offence is a repeated occurrence, any allegation that information is or was being concealed and whether any remedial or preventive action was taken by the alleged offender.[160] According to Environment Canada, having a good compliance history is one of the most important criteria that enforcement personnel will consider when determining if an EPAM is to be negotiated.[161] Presumably this is in recognition of the fact that an offender with a good compliance record will be more likely to adhere to the terms and conditions of any negotiated EPAM. Furthermore, in order to ensure that negotiations for EPAMs are not open-ended and are resolved in a timely manner, they must be completed within 180 days after the Crown Prosecutor has given initial disclosure of the Crown's evidence.[162]

Officially, the key player in the EPAM process is the Attorney-General, in consultation with the Ministry of the Environment. In practical terms, the Crown Prosecutor involved with the case (as an agent of the Attorney-General) and Environment Canada enforcement personnel determine if a given offender meets the conditions for an EPAM. The final decision as to whether or not an EPAM will in fact be negotiated with the alleged offender is made by the Attorney-General or Crown Prosecutor.

EPAMs are flexible and can be used to deliver a number of goals such as retribution, deterrence, rehabilitation and remediation. Section 298 of CEPA states that an agreement may contain any terms and conditions including those which have any or all of the effects set out in section 291. Section 291 lists the types of

[155] See s 296(1)(c) and (i) respectively.
[156] CEPA s 296(1)(e).
[157] CEPA s 296(2)(b).
[158] For an overview of the purpose of enforceable undertakings see s E i previously.
[159] S 296(2)(d)(i).
[160] S 296(2)(d)(ii)–(v).
[161] CEPA Factsheet, 'The New CEPA and Environmental Protection Alternative Measures'.
[162] S 296(2)(h).

court orders which can be made consequent on criminal conviction (as discussed previously) such as adverse publicity orders, environmental project orders, restoration orders and injunctions. An EPAM may also incorporate conditions relating to reasonable costs associated with supervising and verifying compliance with the agreement. An agreement continues in force for the period specified in the agreement, up to a maximum three years.[163] All EPAMs are registered with the court as a public document and are included in Environment Canada's Environmental Registry.[164] Once an agreement is filed with the court, proceedings are stayed or adjourned until a specified time. If the EPAM is complied with (the court must be satisfied, 'on the balance of probabilities'), the court dismisses the charges completely. However, failure to comply with an EPAM is an offence under CEPA and the accused will be prosecuted.

To date, it would seem that six EPAMs have been agreed, with the following two examples provided to illustrate their flexibility and the extent to which they can be used to further a number of enforcement goals. The first EPAM was agreed between Sherritt International Corporation and a representative of the Attorney-General of Canada in October 2001.[165] The defendant company was alleged to have unlawfully exported from Canada a controlled substance, contrary to regulation 7 of the Ozone-depleting Substances Regulations 1998. The EPAM reflected the non-financial aspects of the offence and increased the deterrence impact of the penalty through the use of adverse publicity and community service. It required the company, inter alia to contribute $30,000 to the Environmental Damages Fund, develop a training programme on the export and import of ozone-depleting substances and submit for publication in a journal, an article or paid advertising space describing the facts of the case, the environmental issues raised by ozone-depleting substances and the essential terms of the EPAM. A second good example is the EPAM, agreed in September 2002 and completed in October 2003 between the Attorney-General and Johnson Controls Ltd.[166] The company was charged with two offences relating to the storage and disposal of PCB material and one offence of failing to report a potential release of a toxic substance to an enforcement officer. Under the negotiated EPAM, the company agreed to contribute $40,000 to the Environmental Damages Fund, develop and incorporate into its environmental management system a standard operating procedure for handling ozone-depleting substances and PCBs, develop a training programme for all employees and contractors and publish an article in an approved magazine to alert others to the law relating to PCBs. Upon completion of the EPAM in October 2003, the charges were dismissed.

As is apparent from the two examples above, EPAMs are an extremely flexible enforcement tool which can incorporate a diverse range of enforcement

[163] S 299.

[164] Ss 300(1), 301.

[165] See CEPA Environmental Registry for details. (available at <http://www.ec.gc.ca/CEPARegistry/enforcement/sherritt_agree.cfm> accessed 7 January 2009).

[166] See CEPA Environmental Registry for details (available at <http://www.ec.gc.ca/CEPARegistry/enforcement/EPAMs.cfm> accessed 5 January 2009).

'mechanisms' such as audits, adverse publicity and monetary penalties. All these mechanisms have strong deterrent effects, both independently and collectively. Their being negotiated is undoubtedly a strength as it will increase the likelihood of compliance. However, it also weakens its deterrent impact as regulated entities will be aware of the fact that they have to agree to the terms and conditions of an EPAM before it can be registered.

IV Conclusion

This chapter has examined the diverse range of administrative sanctions available to regulators in enforcing pollution control regulation. Fixed and variable monetary administrative penalties and administrative notices are undoubtedly the most commonly used sanctions, with enforceable undertakings and environmental audits representing new and innovative ways of tackling non-compliance. All these sanctions undoubtedly serve a deterrence function. They all impose costs (either directly or indirectly) on the offender and as such, their deterrent effect can be assessed using the deterrence framework advanced in Chapter 2. Probably the most important factor distinguishing administrative from criminal sanctions is that an administrative sanctioning process is much less costly and time-consuming. This raises the probability of their imposition (e in the framework). The nature of the sanction (F) is the second determining factor in the deterrence framework. With this in mind, administrative sanctions obviously do not carry the moral stigma associated with a criminal prosecution. This arguably weakens their deterrent effect. In addition, there are some administrative sanctions which will probably be of a lower value than any imposed by a criminal court. This is most likely to be true in the case of fixed MAPs although these are only suitable only for minor infractions. But some administrative sanctions are not necessarily less effective in deterring offenders than criminal ones: variable MAPs can be extremely high, administrative notices can restrain or require an offender to carry out a number of acts, enforceable undertakings can incorporate a range of measures. In addition, like many of the creative sentencing options discussed in Chapter 7, deterrence is not the only objective of administrative sanctions. They can identify the reasons for non-compliance and force a change in operational and management procedure, require the remediation or prevention of harm and even suspend or prohibit previously authorised activities. Enforceable undertakings, as negotiated settlements, are even more adaptable and can deliver valuable measures relating to community service and compensation. The flexibility of non-financial administrative sanctions makes them even more influential in bringing regulated entities into compliance.

All in all, administrative sanctions can send very strong deterrence signals. Their cost-effectiveness, flexibility and deterrent characteristics make them an absolutely crucial component of the enforcement toolbox.

9

Conclusion

IN SEPTEMBER 2002, Robert James, his son Simon and daughter Nicola Alger appeared before Cardiff Crown court charged with 10 offences under the Environmental Protection Act 1990.[1] Their company, Gilbert Gardens Nursery and Paving Centre was also charged with eight offences. From 1999, and over a period of several years, the defendants operated an unlicensed landfill site and waste transfer station at their property. Thousands of tonnes of building waste were tipped at Gilbert Gardens, without a waste management licence being in force. In finding the defendants guilty, Robert James was fined £525 (£75 for each of the seven offences), his son £150 for two offences and his daughter £75 for one offence. The company was also fined £10 for each of the eight offences. Despite this stark and egregious breach of the law, as a result of which the defendants earned £85,000 in tipping fees, the total fine amounted to a mere £830. Furthermore, the Environment Agency was refused costs, despite the fact that it had spent an estimated £25,000 to mount the prosecution. This case, which is sadly far from exceptional, brings home the often limited impact of the criminal law in ensuring compliance with environmental standards.

The effective public enforcement of pollution control regulation is essential if we are to be successful in minimising the environmental risks posed by industry. In an ideal world, the regulated community would respond to advice and voluntarily comply with its legal obligations. But the reality faced by environmental regulators is very different, with many regulated entities unable or indeed unwilling to meet environmental standards. In order to move these deviant individuals or firms into compliance, it is crucial that regulators adopt enforcement strategies which will not only deter regulatory non-compliance, but will do so in the most cost-effective way. This is especially important when we consider the limited resources available to most (if not all) environmental regulators. But fair and cost-effective enforcement strategies which are capable of achieving their objectives are not easily identified. A number of common characteristics of environmental regulation complicate this task. Regulatory standards may not always be clear, non-compliant behaviour may be difficult to detect, reasons for non-compliance may vary greatly from one regulated entity to another (and sometimes within a regulated entity), and they may respond differently to enforcement strategies and

[1] Cited in M Watson, 'Environmental Offences: The Reality of Environmental Crime' (2005) 7 *Environmental Law Review* 190.

mechanisms. These factors all go some way to explaining the complexity of the enforcement process and perhaps give an indication as to why the application of law and policy to this area has attracted so much attention. Despite these difficulties, environmental regulators have to adopt an enforcement approach which they believe will best secure a high level of compliance and the law must provide these regulators with a well-equipped enforcement toolbox which is capable of supporting the regulators' chosen strategy.

One key outcome of this monograph has been the development of a theoretical framework based upon Becker's deterrence model of enforcement.[2] By adopting a positive economic justification for compliance which presumes that the regulated community's main objective is to maximise profits, Becker's model predicts that in deciding whether or not to comply with the law, regulated entities will weigh up the costs (notably the expected penalty) and benefits of non-compliance, the expected penalty being the product of the probability of detection, prosecution and conviction, and the sanction that would result from prosecution. Importantly, this penalty includes not only the payment of the penalty, but also the cost of defending any prosecution, loss of business due to adverse publicity and possible exposure to tortious actions. Becker's approach has however attracted much criticism. For example, by adopting a positive economic justification for compliance it seemingly ignores a range of other reasons why regulated entities do (or do not) comply with the law. Also, by focusing on the role of the criminal law and criminal sanctions, it reflects only a small part of the bigger enforcement picture. But I argue that many of these perceived limitations can be reconciled with the Becker's hypothesis and an economic deterrence framework can in fact be used to generate important insights into how best to achieve cost-effective regulatory enforcement. This is particularly so if the framework is expanded in two important ways. First, I present a framework which not only depicts the role played by criminal procedures (namely prosecution) and criminal sanctions in the enforcement process, but also encompasses alternative enforcement mechanisms such as informal warnings, administrative penalties and civil remedies. There is no particular reason why it should be limited to the criminal process. By expanding the framework in this way, it more accurately represents the complexities of modern-day regulatory enforcement in which officials have at their disposal a range of formal and informal, civil, criminal and administrative enforcement mechanisms. Second, and taking this position one stage further, I develop the framework so as to reflect the fact that environmental regulators may progress through different enforcement options and therefore factor in the likelihood and cost of being caught (such as the costs incurred in complying with an administrative sanction or the costs subsequent to civil litigation brought by the regulatory authority), alongside the likelihood and cost of being prosecuted in the criminal courts.

The deterrence framework most obviously aligns with a deterrence-based approach to enforcement in which regulators will punish non-compliant behav-

[2] For a detailed discussion of Becker's approach and the expanded deterrence framework, see ch 2.

iour by using punitive, formal legal tools such as prosecution.[3] But an analysis of most, if not all, systems of regulatory enforcement would reveal prosecutions to be the exception rather than the rule. This is not surprising and given the resource limitations, is arguably a cost-effective method of enforcement. But it is important that we broaden our understanding of the implications of the expanded deterrence framework by recognising its application to other enforcement strategies. I argue that the success of any enforcement strategy, whether it be conciliatory, responsive, risk-based or targeted, depends on the presence of a credible deterrence threat. In other words, the deterrence framework can deliver valuable insights into the enforcement process even where, as is usually the case, a purely punitive enforcement strategy is rejected.

Despite the institutional variations in pollution control regulation and its enforcement in parts of Australia and Canada, and England and Wales, 'command and control' regulation is at the core of the legislative framework.[4] Prior approval (secured by the issuing of a licence or permit) is required for certain regulated activities, and operations are controlled through the use of licence conditions which incorporate inter alia environmental standards and operating requirements. The regulatory controls are underpinned by criminal liability and, in addition, regulators can use a variety of compliance and enforcement mechanisms in responding to violations. Environmental enforcement within these jurisdictions is indeed characterised by its emphasis on encouraging and coaxing compliance; there is a general unwillingness on the part of the regulator to instigate criminal proceedings where an offence is committed. Formal, mandatory enforcement mechanisms such as criminal prosecution and licence revocation are seen more as a 'last resort' and often signify a breakdown in communication between the two parties. This is not to suggest that these regulators never use formal enforcement tools. As we have seen, all of them do issue warnings, do serve administrative enforcement notices and do prosecute for environmental offences. But the use of such instruments is rare and in most cases, regulators appear to resort to more informal methods in order to coax a regulated entity into compliance. But where this less formalistic approach does not achieve compliance, regulators can take formal action in an attempt to secure compliance and deter future law-breaking. Formal enforcement tools must be available if regulators are to advance regulatory compliance.

A key question for us then is how the deterrent impact of enforcement action can be strengthened, given the limited resources that enforcement agencies have to work with. By focusing on expected penalties, deterrence is sensitive to a range of factors including inspection, detection and prosecution rates and the size and impact of the administrative penalty or, where relevant, court-based sanction. There are therefore many ways in which the threat can be strengthened. Violations could be made more visible through better rules, or inspections could be targeted so as to improve levels of detection. I focus on the role that formal procedures and

[3] For a discussion of enforcement strategies and styles, see ch 3.

[4] See in particular chs 4 and 5.

penalties (civil, criminal and administrative) can play in improving deterrence. In all cases, it is imperative that enforcement mechanisms are used appropriately and that the regulated community is aware of how and when they may be put into practice. Guidance is therefore needed to ensure that both regulators and, where relevant, the courts understand the nature of the sanctions and their suitability in particular contexts. It can also promote key values such as fairness, transparency, consistency in application and accountability.

I propose a range of possible improvements to the formal enforcement options available to regulators. For example, bearing in mind that criminal prosecution is probably the least cost-effective enforcement response, regulators may be more inclined to take formal enforcement action (whether criminal or indeed civil or administrative) where proving the offence is simplified.[5] Under the expanded deterrence framework, increasing the probably of criminal prosecution raises the expected punishment in the case of apprehension. One way of easing the prosecutorial burden is to utilise strict as opposed to fault-based liability. This, however, has to be balanced against the perceived inadequacy of judicial responses to environmental crimes. Ironically, low fines are partly explained by the fact that where proof of negligence, intent, etc, is not required for conviction, these offences are viewed as regulatory as opposed to 'real' crimes. This leads me to conclude that a 'tiered' system of liability may be most appropriate, where regulators can select, depending on available evidence, whether to prosecute on the basis of strict or fault-based liability. Staying with criminal sanctions, the courts may feel empowered to impose more effective punishments if they have a range of criminal sanctions which go above and beyond traditional penalties such as fines and custodial sentences.[6] Even if fines were to increase, there would still be instances where they (or indeed custodial sentences) would be inappropriate and ineffective. Alternative sentencing tools such as adverse publicity orders and environmental service orders may strengthen deterrence by not only increasing the impact of any penalty on an offender, but also by improving the likelihood of prosecution or at least creating the impression that prosecution is a more likely response. In many cases, these penalties also have incidental environmental benefits, such as the remediation of harm or the financing of local environmental projects, reflecting alternative penological approaches to sentencing. In Australia and Canada, legislative bodies and the courts have embraced the use of these alternative sentencing options and there are clear moves for a broadening of the judicial sanctioning toolkit in England and Wales. Indirect sanctions imposed using the criminal process can also be extremely powerful in deterring non-compliant behaviour. For example, targeting prosecutions at directors and other high-level corporate executives can be an effective way of enhancing corporate compliance. This, however, should be balanced against the difficulties in piercing the corporate veil and attaching liability to an individual within a company.

[5] See generally ch 6.
[6] See generally ch 7.

Another option would be to improve the range of administratively-imposed formal enforcement tools.[7] The strength of administrative enforcement is that it provides a more cost-effective and speedy response to regulatory violations than criminal prosecution. There are, of course, costs associated with utilising administrative tools such as enforcement notices or licence suspension. The power to issue such directions is not unfettered and due process values such as procedural fairness and consistency of application must be complied with. Ensuring the availability of appeal or review rights is also important, especially in light of the fact that non-compliance with an administrative enforcement tool attracts criminal liability. Despite these procedural requirements, there is no doubt that administrative enforcement is less costly for regulators than criminal prosecution, or indeed any court-imposed penalty. Whilst regulators can usefully employ 'traditional' administrative mechanisms such as notices and licence revocation or suspension, monetary administrative penalties, both fixed and flexible, are a useful addition to the regulatory toolbox and again, are increasingly being utilised in the pollution control context. The recently passed Regulatory Enforcement and Sanctions Act 2008 in England and Wales is a good example of how legislative bodies are recognising the role that financial administrative penalties and enforcement notices can play in the regulatory enforcement process.

In conclusion, the message conveyed in this monograph is not one of more deterrence and more punishment. Of course, penalties should not be so severe as to dissuade people from engaging in productive and economically beneficial activities. It is, however, important to find cost-effective processes and sanctions that will maximise the deterrent impact of enforcement. It is with this in mind, that the expanded deterrence framework can teach us valuable lessons about how to enforce.

[7] See generally ch 8.

INDEX

Index

Index

www.ingramcontent.com/pod-product-compliance
Ingram Content Group UK Ltd.
Pitfield, Milton Keynes, MK11 3LW, UK
UKHW020246290926
15470UKWH00002B/764

9 781841 1392